AUSTRALIANS AT WAR
IN THE AIR

AUSTRALIANS AT WAR
IN THE AIR
1939-1945

VOLUME TWO

- COASTAL COMMAND • THE FLEET AIR ARM •
MIDDLE EAST & MEDITERRANEAN• SOUTH-EAST ASIA
• SOUTH-WEST PACIFIC • GROUND STAFF
• THE TACTICAL AIR FORCE

ROSS A. PEARSON

Kangaroo Press

Author, Ross Pearson, in uniform as sergeant
wireless air gunner.

Cover art by Jim Turner depicting a Beaufighter of 30 Squadron
engaged in a low level 'barge sweep' searching for boats or barges
hidden in the undergrowth above the high water mark.

Cover design by Darian Causby

First published in 1995 by Kangaroo Press Pty Ltd
3 Whitehall Road Kenthurst NSW 2156 Australia
P.O. Box 6125 Dural Delivery Centre NSW 2158
Printed in Hong Kong through Colorcraft Ltd.

ISBN 0 86417 709 7

C O N T E N T S

INTRODUCTION

This is the second volume of anecdotes about air force life—and death—during World War II.

From time to time, but especially at reunions of ex-air force men, I had heard story after story of the experiences of ordinary young men caught up in the dreadful business of war. I felt that these stories warranted a wider audience. I started to collect them and, in the quieter atmosphere of their own homes, the tellers recounted in greater detail what had happened to them. In some instances I was given access to contemporary diaries or written memoirs.

The first volume, placed in a framework of historical background and explanatory material, covered territory familiar to me from my service as a wireless air gunner in Bomber Command in the European theatre. However, I still had a great amount of unused material from other areas of operations and other commands.

This second volume uses that material and against backdrops of the sea, jungle and desert gives additional glimpses of air force life in the war of 1939-45.

This book would never have proceeded beyond the germ of an idea without a great deal of help.

Encouragement to proceed was given by my wife, Dorothy, and by Christine Mansell, the then editor-in-residence, Royal Australian Historical Society. My friend Jack Stronach kept urging me on when I felt the task too daunting.

The heavy job of providing linking material, presenting the stories in an interesting format and correcting the many errors of grammar and logic was made easy by the criticism, revision, redirection and rearrangement suggested by my brother, KW Pearson, AO, former Director of the Australian War Memorial.

The painful work of typing the original format was laboriously undertaken by my wife. The layout, committing to computer files and technical editing was completed by my friend of sixty years, Alan Stutter, DFC, whose comments were much appreciated, and most often adopted.

I would also like to gratefully acknowledge the permission given by the Australian War Memorial to reproduce details of aircraft performance, armament and crewing from the publication *Air War Against Germany and Italy, 1939-1943*. The permission to reproduce material from the Murdoch Sound Archive, namely excerpts from Leo Allen's narrative, is also appreciated.

I also gratefully acknowledge the provision of photographs from the following sources: Dennis Hornsey Collection, Fred Cassidy, Doug Nicol, Bill Hughes, Gerry Judd, Syd Kildea, Reg Torrington, Jack Stronach, Leo Allen, Ron Turner, Doug Moffat, David Nesbitt, Bob Hartman, Bruce Edenborough, Jim Savage, Ron Barker, Reg Hackshall, Laurie Jones, Jack Robertson and Ernie Stanton (RAAF Association), John Appleton and David Innes (author of *Beaufighters over Burma*).

ROSS A. PEARSON
April 1995

1

COASTAL COMMAND AND THE FLEET AIR ARM

The Coastal Command of the RAF, to which many Australians were posted during the war, did not achieve the 'glamour' of the Bomber Command with the latter's thousand-bomber raids and deliberate morale boosting publicity. Nevertheless Coastal Command's activities were in no way less important and in scope were far more wide ranging (in more senses than one) than those of the heavy bombers.

In this chapter we will look at the type of operations conducted by Coastal Command, their importance and their hazards.

Some Australians served with the Fleet Air Arm, and their contribution receives mention in this chapter also.

THE BATTLE OF THE ATLANTIC

Winston Churchill said after the war, 'The Battle of the Atlantic was the dominating factor throughout the War. Never for one moment could we forget that everything happening elsewhere, on land, at sea, or in the air, depended ultimately on its outcome.' Britain's war effort and, indeed, survival, depended on a continual inflow of fuel and raw materials from overseas.

At the outbreak of war the RAF's Coastal Command was ill-equipped for anti-submarine warfare. The use of airpower to support the Royal Navy had been limited to a great degree to considerations of protecting merchant vessels from surface raiders. The role was to be one of reconnaissance.

Thus there were few aircrew trained for anti-submarine work, planes were limited in range and bombs (many of them only 25 pounds) were of little value against a subma-rine. The aircraft available were Avro Ansons with a range of 600 miles, and two squadrons of Sunderland flying boats.

It now seems that the main result of patrols was to ensure that U-boats remained submerged when patrols appeared. The Australian War Memorial's **Air War against Germany 1939–45** *(p. 12) comments that there were even 'scarecrow flights of Tiger Moths,' so desperate was the situation.*

There were some unanticipated hazards as Leading Aircraftsman Les Mills comments. He served at Wick at this time, on 1693 Flight—Anson aircraft:

Seamen in convoys tended to be 'trigger happy' whenever aircraft approached them. I remember when one of our Ansons was going on escort duty, it arrived at its convoy and found that it didn't have the colours of the day. Everything was thrown at it.

The sailors were not much good at aircraft recognition and, fortunately for the Anson, which made a hurried exit, not too good as marksmen either.

Coastal Command became more effective by the end of 1939 and early 1940. More Sunderland squadrons, now fully equipped and operational (including the RAAF's No 10 Squadron) strengthened the Command. Five Hudson squadrons, with a 570-mile radius of action were also operating and the 250-pound bomb was proving a moderately effective weapon.

When, in 1940, Norway and Denmark were overrun and France fell, the German Navy had a choice of many occupied ports as bases. U-boats could be serviced much nearer the scene of their operations and could operate more frequently over a longer range. They were also able to attack shipping where no Coastal Command aircraft could reach them.

When aircraft provided escort, and this was confined mainly to the daylight hours, the U-boats seldom attacked. Instead, they concentrated on night attacks, often attacking on the surface rather than from beneath it.

Both 1940 and the summer of 1941 were the 'happy times' for the U-boats. There was still a lack of sufficient adequate aircraft for effective air cover. The needs of Bomber Command for aircraft were considered to merit greater priority.

Liberator aircraft with their long-range capabilities became available to Coastal Command late in 1941. This made possible the provision of escorts for distances far in excess of what was previously possible. Additionally, a depth charge which exploded at a depth of twenty-five feet was now available and this helped to restrict the U-boats' effectiveness.

The introduction of the Leigh light early in 1942 meant that the cover of darkness did not always protect the U-boat from aircraft attack.

The Leigh light was a lightweight compact searchlight carried by an aircraft and used in conjunction with airborne radar. It allowed an aircraft to approach a surfaced boat at high speed and illuminate it at the last moment to allow accurate dropping of a depth charge.

Jack Robertson explains, from personal experience, how the Leigh light was used:

Provided you were very careful, I thought it was a very good weapon, although it made the aircraft somewhat difficult to fly at the low altitude you had to use in the attack.

When you got a contact you positioned yourself about seven miles away from it at 1000 feet and then homed on it, flying by instruments with the second pilot in the nose ready to lower the light. You would aim to be at fifty feet at one mile from the contact. Height was measured by radio altimeter which showed a green light if you were at fifty feet, the optimum height for dropping the depth charge, a red light if you were below that and an amber light if you were above about fifty-five feet. In any swell, and there always seemed to be one in the Atlantic, you got a series of green to red to green to amber to green . . . which created some tension.

Then, to add to the excitement, when you lowered the light for the final run in—it was hydraulically operated—it hung under the belly of the aircraft and made it longitudinally unstable.

The pièce de résistance was the natural tendency for the pilot to lift his head and fly visually at the target. This was fatal. The pilot was blinded by the beam, lost orientation and, with an unstable aircraft at fifty feet, simply went in. A friend of mine did just that.

The pilot saw nothing of the attack. He simply kept his eyes on the instruments and the second pilot gave the directions, controlled the light and dropped the depth charges. Frankly, I preferred the old Marauder at fifty feet in daylight to anything at fifty feet at night.

The training for this was barely adequate. It consisted of going out and lighting up RN trawlers at night so you could get used to setting up approaches.

By the end of 1942 the Allied air campaign against the U-boats had intensified. Liberators were used in greater numbers in the Atlantic and Catalina flying boats had joined in providing cover for the Gibraltar convoys.

Thus, in 1943, it was necessary for U-boats to stay submerged by day. However, the need to return to base for supplies was overcome by use of supply submarines [milch cows] which replenished individual submarines at a rendezvous in mid-Atlantic. But the combination of aircover from escort carriers and their long-range land-based colleagues helped reduce the effectiveness of the milch cows.

The Coastal Command patrols in the Bay of Biscay were now a significant problem for the U-boat commanders. To meet it, U-boats were equipped with anti-aircraft armament and began travelling on the surface in groups to give mutual anti-aircraft support in case of air attack. It was a hazardous and nerve-wracking experience for a crew attacking, at low level, a small fleet of U-boats all shooting with multiple machine-guns and cannon.

A number of classic battles were fought between groups of U-boats and groups of aircraft with significant losses on both sides.

The U-boats now had increasing numbers of escort vessels and planes to contend with and their successes were limited. In May 1943 for example, 41 German submarines were sunk.

So, by the end of May 1943, the scales had tipped in the Allies favour and air operations in the Bay of Biscay were taking an increasing toll of German U-boats.

North-Western Approaches

Much shipping—many convoys—approached the United Kingdom by sailing round the north of Ireland and down the Irish Sea. Coastal Command vested responsibility for these waters in 15 Group which had bases in Northern Ireland.

Jim Savage served with both 10 and 461 Squadrons RAAF with the rank of LAC and, later, as corporal fitter IIA / air gunner. He wore a gunner's wing because he manned a turret, among other duties, on a Sunderland flying boat. He gives an idea of what convoy patrol was like:

We escorted some whopping big convoys, thirty or forty merchant ships, either coming from or going to America. We'd pick up those coming from America well out in the Atlantic. When we were with the Northern Approaches convoys we'd take them almost into Liverpool.

Sunderland Mark III.

David Corthorn was a member of 292 Squadron and also served in Northern Ireland:

There were four in my crew—myself as pilot, a navigator, a wireless operator and an air gunner.

The nature of the flying varied considerably. It was air sea rescue and at other times we carried a torpedo. With the latter we searched for submarines, trying to catch them when they came up to recharge their batteries. They had to do this periodically—there were no nuclear submarines in those days.

The German submarines would recharge in the Allied sanctuary in the North Sea. The 'sanctuaries' were areas set aside by the Allies where their own submarines could surface with immunity from Allied attack but the Germans tried to use them for their own boats as well.

We patrolled the sanctuaries to search for and attack any enemy U-boat found there. The attack would be made with the torpedo.

The torpedo was a very expensive and sensitive device. It was valued at £25 000 sterling and had to be dropped at a precise airspeed and height (100 knots at sixty feet) if it was to function properly. It also required that the aircraft fly straight and level for some time before dropping to allow careful aiming.

Flying in a straight line at such low speed and low height could be quite unhealthy in the presence of U-boats since they were normally heavily armed with anti-aircraft weapons for just such occasions.

A number of incidents indicated that some crews were not complying with these flight limitations.

To meet this situation the powers that be declared such failures a court-martial offence. To back up this threat cameras were installed to photograph the pilot's instrument panel at the time of dropping to give clear evidence of any exuberant flying at the time of dropping.

My worst patrol was from Portrush in Northern Ireland. We were looking for U-boats coming up the west coast of Ireland. The Prime Minister of Eire, which was neutral, was pro-Axis. Sometimes, when we went down to Dublin on leave—always in civvies—we would see Germans, Spanish and Italians there in their uniforms.

Our Intelligence Officers would brief us, before a patrol, about any Irish vessels reported in the area. We would know what sort of vessels they were and their track. We had to check what they were doing.

One night we were flying just west of the Shetlands. We were relying on our radar and suddenly had a contact.

I said to the W/Op, 'There shouldn't be anything out there'. He said, 'I think it's a ship.' I said, 'Yes,' but there were no vessels reported to be in the area. We turned toward the contact but just before the radar blip got to the centre of the screen the aircraft hit something. It was the cable of a barrage balloon apparently attached to another ship which hadn't shown up on our radar. There were cutting devices fitted in the leading edge of the wings for just this situation and fortunately they worked.

I said to Charlie, ' Our target must be near another ship which we haven't picked up.' So we circled round and came down to have a look. Sure enough, there it was, just visible. It carried the Eire Neutral mark and all the crew went below deck.

The wireless operator had forgotten to pick up a code book when we left base so we notified Scapa Flow in plain language, giving the position of the ship.

When we got back to base the Intelligence Officer made no comment except to ask why we had had no code book.

We heard no more until many months later when the CO received a cable to the effect that Eire had complained that a RN destroyer had sunk a neutral ship. I think this ship was close to where we had found our ship. Obviously it had been supplying German submarines.

South-Western Approaches

Here I have considered the stories of those who operated in an area south of Ireland and far out into the Atlantic as well as the St George's Channel area between Ireland and Wales.

Jim Savage:
When we joined convoys in the south we'd leave them at the entrance to the Irish Sea. Whether they went to Liverpool or to Bristol I don't know. I was only nineteen and I didn't bother myself much where they were going. I only knew we'd be out a long time and I was going to be very tired. All I wanted to do was see Wolf Rock and the Scillies coming up and know we were close to home.

The number of escorts were, initially, very very few. When I started flying my first convoy trips in 1941 the sea escorts were almost non-existent. We were the only air escort—one Sunderland circling the convoy hour after hour until another Sunderland came out. Nevertheless convoys loved having us and when we came low they'd always wave to us.

Navigator Doug Moffat was on a Sunderland sent to escort a convoy. He describes what happened:
We didn't ever find the convoy—mainly because the weather was down on the deck. The position we were given might have been right or wrong. We did find a poor little corvette slopping about in the drink. It was obviously lost from the convoy and it flashed us a WAI (Where am I?). We said, in effect, 'We'll bite'.

When we got back to base the question of the missing convoy was gone into. It was concluded that the visibility had been so bad that the convoy had just scattered after a U-boat attack.

You could only just see through the mist that time. I only found the corvette because of SE [special equipment]. This had a linear range scale with a range of ninety to one hundred miles and you could home on an object.

While on patrol a watch was also kept for planes that had crashed in the sea, lifeboats etc. Doug Moffat tells of one lucky American aircrew:
We found an American Liberator crew out at sea. They were in the Bristol Channel and we were ordered to stay with them until the Air Sea Rescue craft came out. They were in their dinghy and had been operating just adjacent to our station. They were picked up successfully and we aborted the rest of our patrol.

Bay of Biscay

Although it is part of the North Atlantic Ocean I have chosen to group stories about this area under a separate heading. The Bay of Biscay stretches from Brest (France) virtually to Cape Finisterre in Spain. U-boat bases were located along the coast of Occupied France and saved long voyages to Germany to refuel and refit.

461 and 10 Squadrons were RAAF squadrons and their crews were basically all Australians. There were a few who were not. Jim Savage, a fitter IIA/air gunner, tells of one such and his fate and illustrates the dangers of what might seem a boring chore. Success did not always come to the searcher:
From time to time you come across someone who says he doesn't like the French. If you go to the trouble to question them you almost invariably find the dislike is based only on anecdotal evidence—someone else's supposed experience.

Perhaps I am biased the other way—I have had first-hand experience.

My squadron was a purely Australian permanent RAAF entity operating in 19 Group of the RAF Coastal Command with a home base, at the time, on Plymouth Sound in the south-west of England.

One day a badly damaged, strange-looking aircraft with a large central float and small ones at the tips of the mainplane—which folded like a jacknife—was brought into our hangar. With it came a man well over six feet tall and built in proportion who no-one believed could fit into the tiny cockpit of this kite. He wore a dark blue uniform of naval cut and smallish peaked cap. It turned out he was Adjutant Hazard.

Adjutant was a French rank roughly equivalent to the rank of flight sergeant and he really was the pilot of the frail craft which was carried on the giant French submarine Surcouf. It had escaped the Germans on the fall of France and was now supposedly under Free French control.

Jim Savage and L.A.C. Crew member in flying gear.

Jacques had had a prang was the first story we heard, but the damage had really been caused in getting it back into its mother ship's hangar in poor sea conditions.

It wasn't long before the Sergeants' Mess, crews and ground staff alike, got to know and like this extremely likeable Frenchman with his big broad smile and rather poor English. It wasn't long before he was scrounging flights to break the monotony of waiting round for repairs to be carried out.

As our aircraft was one of the squadron's oldest we were, at that time, mostly doing local jobs—ferrying, bombing practice at Stranraer etc.—and so it was easy for Jacques to get off with us.

The Surcouf was off somewhere in the Atlantic and due to some unknown complication no work was being done on the little aircraft. The weeks stretched on. Jacques acquired not only a RAF uniform with FRANCE on the shoulders but a fair command of English and good Australian bullocky language.

We picked up a new aircraft from Short Brothers at Rochester and were back on proper operational duty. Jacques continued to fly with us. Tom Stokes, our skipper, had visited France with his parents prewar and took a great liking to him. Gradually Jacques became fully qualified to fly as second pilot, then as first pilot and was later good enough to be captain. But that was, of course, impossible in an RAAF squadron.

He continued to wear RAF grey until, one famous day, he fronted up to the CO, Wing Commander Richards, with yet another request to wear the RAAF uniform. Richards asked him why and he gave the oft quoted reply, 'Mr Richards, I will tell you the truth. I am shit to death being taken for a bloody Pongo!' The very next day he was decked out in the blue (still with FRANCE on the shoulders). With that and his wings he was a very proud adopted Australian.

Now I was always a rather cheeky bugger and I used to get a lot of fun out of giving Jacques a good ribbing. One day he was the pilot with two crew members—of which I was one—on storm guard. The idea was that there would be a fitter IIE/AG to start the two outer motors and a fitter IIA/AG to have the moorings properly placed when a big blow came up. The pilot had to keep the strain off the moorings and, if the worst came to the worst and the moorings parted, to taxi. The blow didn't eventuate and I made some outrageous crack at Jacques. He jumped to his feet, grabbed me by the collar and the seat of my pants and banged me up and down on one of the bunks. I remember his words to this day—something like, 'Be careful you little Aussie bastard. One of these days you go too far.' He then stormed off to the bridge. I could see something was bugging him and took no offence and we stayed good friends but I couldn't resist still taking a poke at him from time to time.

The Free French authorities in London were anxious to get him away from 10 Squadron but this proved difficult. He flew on any patrol he could get on and many of their efforts were abortive because he was nowhere to be found. He flew with us much of the time, probably knowing that his time was almost up.

There came the day when our aircraft was 'up on deck' and I went along to the stores to exchange some worn gear. Jacques was there and, needless to say, we started talking about the impossibility of him being able to stay with the squadron—particularly as captain. I said to him, 'Well Jacques, you big French frog, your Free French mates will get you soon and slap you in the RAF grey again.' and he said, 'I will tell you something Jimmy. They will never do that. I will always wear the Aussie blue.' and we parted—me to a night out in Plymouth and he on a scrounged ASR [Air Sea Rescue] flight down the Bay.

He never came back. Neither did one of my best friends, also in that crew. I didn't find out until after the war that they were jumped by a lousy Arado 196 which must have come hurtling down out of the sun. It had cannon—they only had Browning .303 and a few lousy VGOs (Vickers Gas Operated machine-guns). Jacques was, of course, wearing the Australia blue.

On my first visit to France after the war I managed to track down his mother and sister at Soissons. I went up from Paris with a handful of photos and other notes on Jacques. They only had a few Red Cross notes from him or about him. So I stayed with them and we talked. Madame Hazard told me that one of Jacques' great heroes was the navigator La Pérouse and how much he would have loved to be in his crew and see Terra Australis.

Well, he didn't. But he was as much an Australian as any one of us and always will be—in the Aussie blue of course.

Jim Savage mentions an encounter:

A Focke-Wulf Condor came in pretty close to us on a convoy trip. We fired at him. I don't know whether we hit him but he sheered off very fast. Of course that was the idea—we were there to convoy the ships, not to go chasing after Condors. But it was nice to know he had gone home because he was a spotter for the U-boats.

Doug Moffat relates a story about his squadron:

The official histories of the Royal Air Force and Royal Australian Air force in World War II relate the story of Sunderland N of 461 Squadron RAAF. This aircraft and its crew, captained by F/Lt Col Walker, achieved what had never been done before.

While on a routine anti-submarine patrol in the Bay of Biscay N/461 was attacked by a formation of eight Junkers 88 twin-engined high-performance fighter aircraft. The time was late in the afternoon, the sky was cloudless and N/461 had no option but to fight. This N did—very successfully.

N/461 shot down three of the Ju-88s, probably destroyed two more and damaged three so severely that only two returned to base. This was subsequently verified from radio-listening sources.

N/461 returned to Britain as a battered, tattered and shattered wreck of a flying boat which was forced to alight off a beach on the southern coast of Cornwall.

The crew managed to wade ashore from the beached aircraft carrying their dead and wounded with them.

Col Walker could see in the darkness of the night a blacked-out coastal marine station on an adjacent headland. He promptly climbed up there and requested use of their telephone.

It was now late at night. The telephone rang on the bar of the Officers' Mess. The steward answered it and passed it to the CO of 461 Squadron.

'Who is it?'
'Walker.'
'Walker? You're dead. Where the hell are you Walker?'
'Penzance.'
'What are you damned well doing at Penzance?'

At this Col Walker's calm and cool which had sustained him and his crew for the last twelve hours deserted him completely.

'Playing at pirates you silly bastard.'

Doug Moffat had an experience of mechanical failure while on patrol:

I went to 10 Squadron as a pilot officer and flew with a Sergeant Strath who was captain. Our first pilot was a warrant officer and I was the only commissioned officer in the crew at the time.

In the winter of 1942/43 there was snow in southern England but not at Mount Batten, which was our base. Bing Crosby was singing 'I'm dreaming of a white Christmas' but having just come from Rhodesia I didn't appreciate the change in temperature.

My log book shows I carried out 21 operations on 10 Squadron—most of them with Sergeant Strath who rose rather rapidly to be first a pilot officer and then jumped to flight lieutenant. It was felt that at least pilots and navigators should be commissioned officers.

My first trip was labelled anti-submarine patrol and I have the notation 'Duration five hours—duty not carried out.' Most of our jobs were about twelve hours although the aircraft was supposed to have a sixteen-hour endurance. The squadron reckoned on twelve effective hours because of the likelihood of diversion.

We lost an engine on this trip. The Mark I and Mark II aircraft had Pegasus Mark 18 engines, which meant the propeller could not be feathered if the engine failed. This aircraft was one of those. The problem then was that the propeller on the dead engine windmilled and windmilling props were liable to come off. If it flew off it was an interesting question as to whether it would come inboard or go outboard.

On this trip we had to fly for two hours with a windmilling prop which, fortunately, stayed on.

My pilot on this trip was Flying Officer Griffiths. You can tell from his rank that he was not far up the scale. To be a full captain with the rank of flight lieutenant you had to have 500 operational hours on Sunderlands.

A tour was close to 1000 hours on 10 Squadron. It was 800 hours or seventy missions, whichever came first, on 461 Squadron.

I had a sergeant pilot because 10 Squadron was RAAF and many of the aircrew up to this time had been permanent staff and you could have a straight gunner with the rank of LAC.

An interesting result of this approach was that sergeants got flight pay and Sergeant Keily DFM got his medal although he didn't have a flight engineer's brevet (half wing).

Jim Savage relates an incident of particularly bad weather:

One of the first crews I was part of had an old permanent Air Force fitter IIA/air gunner named Jim Burnham as first fitter. This meant he was the crew chief (later called flight engineer) and my superior officer.

I was crewed as the rear or tail gunner and for the next couple of months he treated me the same as the others—perhaps a little less warm and friendly—but that was to be expected because he had flown with them long before I joined the crew. After all, I was just a kid of nineteen and, I would have to admit, a little brash. I didn't notice any change until after a trip to Gibraltar

The weather in the Bay of Biscay had been atrocious and didn't soften any until we were well south of Cape Finisterre. Nearly all the crew were horribly airsick. Funnily enough I, who had been sick in everything that moved when I was a kid, didn't feel any discomfort at all.

Jim Burnham sat at the bench for the whole flight sucking at his pipe. The second fitter was completely out of commission—as was the skipper for a fair while, one of the wireless operators and the armourer. In such conditions it was impossible to maintain full gun watches—and really quite unnecessary—so those who were fit filled in wherever possible. Hot drinks seemed to help so I kept the galley running.

When we eventually moored up in the harbour at Gibraltar, one of the first things Jim did was call me aside and thank me for doing 'so much to boost the crew's morale.' It bucked me up no end.

Jim Savage saw boredom as a problem:

The biggest problem was boredom—the trips were twelve to fifteen hours. We were not assigned to any particular turret and used to swap around. We always kept the mid-upper and rear turrets manned but only manned the nose turret when in action. Whoever was doing a stint in the galley had to grab his helmet and race up to the nose turret as soon as the klaxon went.

We used to have a kip—legitimately—in the bomb room—where the bombs were kept. They were carried internally in the Sunderland. When wanted, the doors dropped down and the bombs were swung out—they would have slowed us down about twenty-five knots otherwise.

Doug Moffat takes us on a patrol:

Normally you'd be wakened by the duty officer at dawn. You then had about half an hour before you adjourned to the aircrew mess for the Last Supper or early breakfast.

The captain, navigator and senior wireless operator normally a wop/ag) would be briefed on the patrol by the Station Intelligence Officer and the Met people. They would then board the aircraft.

Meanwhile the rest of the crew would have gone out to the aircraft by dinghy. They got the aircraft tied up on short slip ready to slip the moorings. Motors were not started until then.

About two hours would have passed.

We then took off into the wild blue yonder and got to cruising height well before the datum point—48°N, 10°W in most cases. This was both the beginning and the end of the patrol. The patrol lasted about ten hours but this didn't stop you diverting if you saw something worth investigating. Anything unusual had to be reported.

While on a patrol with 461 Squadron we had been sent to search for a submarine down near Spain. We arrived at the point where it had been seen and decided to do a creeping line search towards the west. This was really a rectangular zigzag with the zigs decided by the visibility. Sure enough, after a while there was a swirl in the water ahead and we ran the bombs out. We sent off a 465 message—it means, 'I am attacking a submarine'—and prepared to drop depth charges. Suddenly it surfaced—it was a whale and was bang on the position of the alleged submarine. We didn't drop the depth charges.

Jim Savage notes:

The squadron directed vessels to people in lifeboats or people from downed aircraft. You could land and take off a Sunderland in a calm sea but it was pretty dangerous if there was any swell at all. The floats could break off in a choppy sea and the aircraft would turn turtle. The Catalina could land with its floats retracted and lower them as it came to a stop but ours were fixed and prone to damage.

Doug Moffat had an unexpected involvement in a rescue:

We were on the way home from a patrol and had reached the POE (point of limit of endurance). We had five hours' fuel left when what should we find but a great swirl of oil with about fifty U-boat survivors in the water.

We reported them. They were the survivors of a Sunderland attack. We managed to home a naval vessel to them, they were picked up and we set course for home after our good deed. It really was not a boy scout effort as they could provide much useful information in the fight against the U-boats.

Three U-boats were sunk that day—it was a great day for the cooperation between 19 Group and the Royal Navy. The fight against the U-boat had come to a crux in the summer of 1943. Up to then the U-boats using the ports in the Bay of Biscay would submerge at the sight of an attacking aircraft. But they then decided to stay on the surface and fight it out. Once a U-boat sighting was reported all aircraft in the vicinity congregated for the attack.

On this occasion, apparently, a group of U-boats formed a circle to protect themselves. Two were sunk by aircraft and the third by a naval sloop.

Doug assisted in another unusual rescue:
We did one run to find some American tank-landing craft which were lost on their way to England for the D-day invasion. They were hopelessly lost. They were flat-bottomed, not very seaworthy and were supposed to be under the guidance of the navy while on their way from America. Somehow they had become separated from their escort in bad weather.

We managed to find them heading towards Brest. We reported their whereabouts to Group and advised them that there were surface vessels on the way to guide them to a friendly rather than a hostile destination.

Doug Moffat had another interesting contact with surface vessels:
We were airborne on Christmas Eve 1943—it was the middle of winter—and proceeding on patrol when we received an order to find an enemy force last reported heading west. We had no idea what the force was so we headed gaily off.

We picked it up on radar—about ten blips. Then, as we came out of cloud, our nose gunner, who was a Yorkshireman and almost unintelligible on the intercom with his accent not helped by the noise of the slipstream from the front turret, sighted and identified ten ships. They were all enemy destroyers coming out from one of the French ports to escort in a blockade runner.

We reported them and shadowed them. Our orders then were to keep shadowing until relieved. We kept out of range of their guns—each had radar directed high-angle 4.5 inch guns. We just sat back and watched the blips on our radar. We had an individual range to each ship on our nose or beam and circled them at about fifteen miles as they went west of north. We were relieved and that was the last we heard of it until Boxing Day.

The weather was lousy that day and aircraft couldn't get off the deck to keep up the watch. We were alerted and were just about to try to get airborne when we were recalled. We weren't too worried about going because the Poms were pretty good at diverting you to a base in the UK where you could get in—or even Gibraltar.

Although we didn't get off they must have got some long-range stuff up because, on the 28th, eleven destroyers and the blockade runner were sighted approaching port. The supply ship was half-full of rubber from Malaya.

A mad Czech pilot in a Liberator attacked, set the blockade runner alight and even managed to get away himself. This was good enough for the *Enterprise*, a 10 000 ton cruiser like the *Canberra*. It out-gunned the destroyers and just about knocked off the lot.

Dinghies sighted by aircraft at Coastal Command.

We often saw fishing boats and we were warned that, while they were supposed to be neutral—French or Spanish—Intelligence reckoned that they carried German spies on board who reported on the frequency of patrol aircraft.

There was still some chivalry in combat. If the German fighters forced a plane to drop its depth charges (we carried eight 250-pound depth charges) they didn't always press home their attack. If the plane was shot down after dropping its charges it was rare for the survivors to be machine-gunned.

The role of the navigator was particularly important in Coastal Command. Doug Moffat tells of the procedures and responsibilities involved:

Navigation was basically dead reckoning (DR) which relied purely on observation of the weather conditions and aircraft drift. Both 10 and 461 Squadrons had a reputation for very good DR.

The Coastal Command drill was three drifts per hour, one three-course wind every hour and a position line per hour—it could be an astro sight or D/F from the receiver mounted above the navigator's position.

We could tune in to all the recognised broadcasting stations given in the transmitter briefings. These stations could be either on the French coast—Radio Calais for example, where Hildegarde sang—or the Spanish coast. The French used to change the sites of their transmitters from time to time so our accuracy was influenced by the accuracy of our intelligence.

We could pick up the Spanish stations but the Spanish were not very cooperative. At one stage they had a very good location beacon near Cape Finisterre which was used by the U-boats and ourselves. But the Germans asked them to move it and they did.

You might well ask how navigators stayed on the ball for twelve hours. In fact this was the great whinge amongst navigators. The crew consisted of the captain, first pilot, second pilot, two wop/ags, two flight engineers and sometimes two straight gunners—but only one navigator. So the navigator was the only one who had to do the whole thing himself.

In the case of fighter attack the navigator became the controller. He took up position in the astrodome and directed the gunners and evasive action. An attacking aircraft generally lined up on the beam, got the pace and developed the attack either on the beam or the quarter.

An attacker on the starboard beam, for instance, would turn to port to intercept your course. The basic tactic for the controller was to turn to starboard, into the attack. As soon as he saw the nose come round he gave the order

'Go!' and, if the aircraft had the height, it dived to starboard. If you saw the attacker had the right deflection you would call, 'Steeper, Steeper'—the only way to go faster was to dive. The controller watched and, if the attacker dived under the tail, he warned the rear gunner.

We had squadron aircraft attacked by six, eight and even by fourteen Ju-88s at once. We rarely patrolled with another aircraft. The whole aim of the anti-submarine work was to patrol rigid lines and this would keep submarines down.

Doug Moffat enlarges on the responsibilities of some other members of the crew:

We had a wardroom, a galley and a bomb bay We carried an anchor in the galley. Generally one of the flight engineers or gunners cooked. The wireless operators were on continuous watch.

Most of the inward signals were in cipher and had to be deciphered from a chart and code. The wireless operators also acted as gunners. They maintained a continuous listening watch and messages on the aircraft's call sign were taken down. As a relief to boredom messages to other call signs were also taken down at times and these sometimes gave us useful information about events elsewhere.

In 1943 the German Admiral Doenitz ordered all U-boats crossing the Bay of Biscay to travel in convoys on the surface.

Doug Moffat comments:

Aircraft had to deliver depth charges from no higher than fifty feet. The captain would line the aircraft up like a naval torpedo attack. He would try to cross over the conning tower at an angle of 45 degrees to the length of the U-boat to give the best chance of sinking or damaging it when they straddled.

However, a U-boat on the surface had a favourite method of responding to this attack. It would turn towards the aircraft to give a narrower fore-and-aft target. In addition it made a non-deflection shot at the aircraft. As the wingspan of a Sunderland was 108 feet it was not a small target.

Despite the difficulties and dangers there were many successful attacks. Although Doug Moffat did not sink a U-boat his Group was very successful, as he indicates:

The aircraft from No 19 Group, to which my two squadrons belonged, sank fifty and damaged fifty-six U-boats out of 2425 passages from Biscay bases up to May 1944.

During forty-one months 19 Group lost 350 aircraft in the Bay. The results would have been better had we been given more good aircraft but they were given to Bomber Command.

Jim Savage comments on the U-boats:

Until 1943 the U-boats would dive when they sighted an aircraft. We liked to get them on the surface as our depth charges were set to explode on impact, but if we saw them on the surface they were usually submerged by the time we got to them.

There was a stopwatch fitted to the controls and the skipper punched it as soon as a U-boat was sighted. If thirty seconds elapsed before we could drop our load we had to circle while the armourer reset the depth charged to explode at a given depth.

Doug Moffat elaborates:

Most of the captains who got U-boats did so by underwater attacks. From the look of the swirl and the elapsed time by the stopwatch the captain would estimate how deep the sub would be and its probable course.

Submarine captains would turn port or starboard depending on the geography. If there was a shoreline close by they would turn away from it. Most hugged the shore going into and leaving the French ports which were under German control and heavily used. They saved the U-boats a long trip around the north to their bases in the Baltic Sea in Germany.

At the time the U-boats did not have radar but had an early warning device which is why the technique of attacking the swirls was developed.

Roy Baxter was a navigator on a Liberator operating in the Bay of Biscay at this time. He tells how his pilot attacked a U-boat:

On my second trip my regular skipper was ill. That night [2 January 1944] we were flying down the Bay of Biscay on anti-sub patrol. We were about half way down the bay when, about midnight, one of the wireless operators picked up blips to the east—towards the French coast. The skipper decided to go and investigate. We found a pack of U-boats on the surface. There were anything from six to ten of them.

We decided to attack with depth charges. We were at a height of 150 feet and about a mile away and didn't use the Leigh light as it was bright moonlight.

We dropped four depth charges on the first boat. The mid upper gunner fired at them and, of course, they fired back. The bigger U-boats carried anything up to a 4 inch gun.

The skipper decided to carry out another attack and use the four remaining depth charges on another boat.

We went down to attack the second one but it disappeared beneath us. Suddenly the Lib went up in the air. I heard a crump at the back and I knew we had been hit. I didn't know how serious it was but we kept flying and the lights went out. The skipper called out

'Anyone hurt?' and we all answered except Geoff Kelly who was the wireless operator and was operating one of the beam guns down behind the bomb bay at the time. It transpired that he didn't feel hit but could feel blood trickling down his face.

The skipper thought Kelly might be in shock and we ought to have a look at him. I went down through the bomb bay to where he was and shone my torch on him. It wasn't blood trickling all over him. It was the contents of the Elsan [the portable toilet carried in the aircraft].

The shell had come up between the bulkhead at the bomb bay (i.e., just behind the depth charges) and the big radar scanner. It was fortunate it did come up there otherwise we would have copped the lot.

The aircraft could still fly so we made our way back to Cornwall. Having made an attack we had to go and see the AOC and give him an account of what had occurred. When we told him about the Elsan it made his day—although Kelly wasn't too welcome.

Baxter was engaged in a second attack on a U-boat:

The occasion was 7 June 1944—the day after D-day. We were flying across the Channel, along the French coast near Brest.

As we went through the strait between an island and the mainland a U-boat surfaced dead ahead of us. We were too close to drop straight down to the attack so we went past and our gunners fired at its gunners who were running to man their guns.

We had to circle round to come in on an attack path at 150 feet. On this trip we were not carrying our normal load of 250-pound depth charges but had a huge acoustic torpedo. This was very hush-hush at the time—only pilots and navigators were briefed on it. The theory was that, when dropped, it homed on to the U-boat's propellers.

We dropped the torpedo but the U-boat had dived and had been submerged about twenty seconds and would have turned away so there was no result for us.

Immediately after dropping, the rear gunner called out 'Two fighters coming up from astern' and the skipper was about to take evasive action and the rear gunner about to open fire when they turned away and we recognised them as Beaufighters, but they looked just like Ju-88s as they came up on us.

Roy Baxter was lucky—he survived. As he says:

Hunting U-boats in the Bay of Biscay was not all beer and skittles. When I was operating in the bay before and after D-day we lost 12 aircraft, half the squadron. Some of them were shot down by fighters because we operated right up to the French coast. Most, though, were shot down by U-boats because we had to attack at low height. All were shot down in the sea and no crew member survived.

THE ARCTIC CONVOYS

The Early Stages

The Allies could not provide troops to assist the Russians in their land operations in Europe. They could, however, provide munitions and supplies.

The logical supply route was by sea around the North Cape of Norway and through the Arctic Sea to Murmansk and, later, Archangel.

All went well until March 1942 when German aircraft and U-boats operating from Norway began attacking the convoys.

The result was a repeat of the Battle of the Atlantic with the added complications of being always within range of shore-based German aircraft and major units of the German fleet stationed in Northern Norway. To this was added the almost perpetual daylight of the far Northern summer and almost perpetual night of the winter with the appalling weather of the latter.

Convoy losses were heavy. The climax came with convoy PQ17, a convoy of thirty-four ships. It was scattered when the cover was withdrawn on news of an impending attack by the heavy units of the German fleet. Twenty-three ships were sunk. Only 70 000 tons of supplies were delivered out of the 200 000 tons shipped. The Admiralty seriously considered suspending the shipments.

However, Sir Philip Joubert, AOC Coastal Command, proposed stationing a force of search and strike aircraft in Russia as a temporary measure during the operation of the next convoy. This was agreed.

Jack Davenport relates a little of what followed:

We did a conversion course to find out how to drop torpedoes from Hampdens. They were normally dropped from Swordfish aircraft at a much slower speed than could be achieved from a Hampden.

We had to find a way to drop them so that they would hit the water and submerge without damage and then run straight. A wooden frame was devised and we dropped them from 120 feet instead of the more usual 60 feet.

It was a fairly chancy and hazardous business. The torpedoes were World War I vintage and would travel at about twenty knots so the chances of hitting a capital ship were pretty slim. The target had time to turn head or tail on and make a narrow target. If the torpedo struck a glancing blow on the hull it may not explode and, in any case, the resulting damage would be reduced.

In September we were sent to Russia to defend what was the largest convoy sent there. I think there were seventy-six ships. The Germans had *Tirpitz*, *Admiral Scheer* and *Hipper* with four destroyers in Alten Fjord in

northern Norway nicely placed to attack the convoy.

We flew to Russia from the Shetland Islands. It was beyond the normal range of our aircraft but by ignoring the normal margin of safety we could just make it.

We lost 25 per cent of our force getting to Vaenga in the Kola Inlet. The Germans shot down a couple, the Swedes got a couple but I don't know what happened to the others. Twenty-three aircraft survived the flight.

I ended up landing at a place called Maechegorsk, which nobody had ever heard of.

I don't think even the Russians knew where it was. It was in the middle of the tundra and that's all it appeared from the air. The Russians had it extremely well camouflaged because it was only about fifteen miles from the front line and was attacked about twice a month by the Germans flying from Finland.

I didn't actually choose our place of arrival. I was flying quietly along when I was attacked by two Hurricanes flown by Russians. We later learned that nobody had told them we were coming. They were very persistent. They kept coming in and having a few shots at us. My rear gunner, who was from Lancashire, was very keen to have a go at them. I managed to stop him firing, we fired off the colours of the day and put our landing wheels down. They formatted on us and pointed down so that's how we came to Maechegorsk.

We had a wonderful party there in the bush, although the camp was as crude as crude could be. There were no roads or anything and we had a breakfast of mushrooms, radishes and other similar things—most unwelcome sort of food. But then they gave us a pannikin of what turned out to be vodka. It wasn't very good vodka. They said it was made from wood pulp. It must have been pretty crook wood! We had a party for several hours.

We kept saying we wanted to get away and they said they couldn't get us a clearance. We eventually found out that they were waiting for a pilot to escort us to our proper destination. In recognition of the fact that we were their staunch allies the escort had to be a pilot who had been awarded the Russian Legion of Honour.

We at last got away after they had filled our tanks from drums labelled 'Kerosina'. That worried us no end but we took off without any problem and flew to our base where we received a great welcome. The Russians were very generous and gave us their barracks. We later learned that they were living in caves up in the hills because the base had been bombed every day.

Our objective was to get the convoy to Russia. The Royal Navy escorted the convoy part of the way and it was left to our aircraft—each with one torpedo and .303 machine-guns—to defend the convoy against the German

Navy, which was perhaps the greatest in the world at that time. The *Tirpitz* was certainly the greatest battleship.

Daily reconnaissances were carried out by Spitfires which had been shipped in. We took off on one occasion to attack the German fleet which had been reported as raising steam to leave the fjord. We needed to intercept the Germans well before they reached the convoy.

We had a long and fruitless search. We searched the Barents Sea and stayed there longer than the scheduled time but didn't see a single enemy ship.

Most of the convoy reached port safely even though they had been attacked by U-boats and aircraft from Northern Norway.

We were supposed to stay in Russia for ten days or a fortnight. In the event we stayed for four and a half months. Because of the distance we had had to fly to get there our personal kit had been limited to one change of underwear and one spare pair of socks.

The Last of the Convoys

Maintenance of the Russian supply line by the Arctic route became unnecessary when it became possible to ship via Persia and the Persian Gulf.

However, there were still empty ships in Russian ports waiting to return to Britain. Also waiting to be brought to Britain were the crew of USS **Milwaukee** *and a Russian crew to take over the* **Royal Sovereign.**

A strong escort was sent from Britain including the aircraft carriers **Activity** *and* **Fencer** *and sixteen destroyers.*

The home-bound convoy consisted of forty-five ships and only one was lost on the journey. Despite strong winds and snowstorms the carrier aircrews flew their Swordfish aircraft and destroyed three U-boats—U277, U674 and U959.

Reg Torrington was an observer in a Swordfish flying from HMS **Fencer.** *He tells of some of the incidents of this voyage:*

I flew off escort carriers in 1944/45. Going to Russia in HMS *Fencer* the weather was cold. The flight deck was covered in snow—it was very thick and we were allowed up on deck to have snow fights before they cleared it.

We really didn't strike any opposition until we got inside the Arctic Circle.

On the way back the Germans were with *Fencer* all the way. The U-boats were very professional.

On the way out, if they knew it was an important convoy they had a reconnaissance plane and they might get on to a convoy early. They used to go out further down from Stavanger or the other side from North Cape or Narvik or Norwegian ports. Generally snow storms were intermittent and there would be no point generally

in sending an aircraft off for two or three hours in a snowstorm because you couldn't see anything.

We went to Polyarnoe, the Russian naval base which was just a speck on the Kola Inlet. The merchant ships went to Murmansk. There would be nothing to do there.

Polyarnoe was very interesting even though there was nothing ashore. I went ashore anyhow.

The Russians were very hospitable. They loved coming on our ships—I think for a change of atmosphere.

In this we were very lucky. I don't know whether it was because we had a great variety of people on board or because we had the concert party on board.

We brought back a Russian admiral and his staff. They had been going to Britain on a Dutch passenger ship but it broke down.

Aircraft navigation was by dead reckoning. We had a chart with the searches entered on it. We flew from the fixed base and then plotted the course and wind to find the track followed.

We watched the wind carefully to be sure we saw a change. If a change was missed the navigation was out and you didn't get back. We noted the changes of wind by watching the water. We flew at 2000 feet and watched the water very carefully. We became very skilled at determining the wind this way.

Sub Lieutenant Reg Torrington, Fleet Air Arm.

HMS *Fencer*. A Fairey Swordfish is taking off from the deck.

I often think back to those times and how I could see where the wind was blowing and find the course from this.

At one stage we had a very difficult commander. We made one of the searches in an Avenger. We went for two hours and when we came back the carrier was not there.

We had to do a square search. Each leg we would keep increasing it. We had to maintain radio silence but finally homed in on the carrier and landed half an hour later. He had altered course by 60 degrees—he was hopeless. We were very lucky to get back.

One of the biggest problems was the incredible cold—particularly in the open cockpit of a Swordfish. I had a flying suit and an inner flying suit which was very warm—it seemed about an inch thick. We also had outer gloves and inner silky gloves, goggles and flying helmet.

When I first started we didn't have intercom—instead we spoke into a tube with a little mouthpiece to communicate with the pilot.

When I transferred to the Avenger I had heating inside and an all-rubber flying suit which fitted tight around the wrists. We wore gauntlets over our hands and a hood over the head. This was supposed to keep you alive longer if we came down in the sea.

I don't know how long one could stay alive in the sea but in 1944 half the crew of a ship were frozen to death after half an hour in the water. They had life rafts which we did not have so our chances were rather less.

We had radar to pick up subs. We went out about 200 miles from the carrier. One plane went about 100 miles ahead of the convoy and flew backwards and forwards maintaining a lookout ahead of the convoy. Another aircraft circled the convoy.

On one submarine approach we got caught in the slipstream of the plane ahead. It felt as though someone had hit the aircraft with a giant sledgehammer.

The Fairey Swordfish — although designed well before the War, it proved itself a useful workhorse for the Fleet Air Arm.

In a torpedo attack the ideal thing is to surround the ship and attack simultaneously from different directions. In this case you are more worried about colliding with other aircraft than the ship's firing. You're always looking round.

We used three aircraft. One would machine-gun the sub, the second would be a Swordfish which fired rockets and the last plane in dropped depth charges.

If the sub was submerged there was not much you could do. You could try dropping depth charges but generally the escort ships took over.

OFFENSIVE CAMPAIGN AGAINST U-BOATS IN THE NORTHERN SEAS

By May 1944 British Coastal Command was in a position to start an offensive against U-boats in Norwegian waters. In June the Allied forces landed in Normandy and two months later began capturing the U-boat ports in the Bay of Biscay. This forced them to move to Norway.

One of the aircraft used by Coastal Command was the Catalina. John Appleton was a member of the crew of a Catalina which sank U-boat 347 in this area on 17th July 1944. He tells his story:

By mid morning the teleprinter from 18 Group Headquarters had ordered our Squadron to provide a Catalina to take off around 1300 hours and proceed to a patrol area roughly bounded by latitudes 68 to 70 deg. North and Longitudes 4 to 6 deg. East. This was an area a couple of hundred miles west of the port of Narvik.

Subsequently a U-boat was sighted and the attack was mounted. John Appleton continues:

The skipper—Cruickshank, a Scot—had manoeuvred into a perfect attacking position astern of the submarine, just keeping out of range of the enemy gunfire. He gave a blast on the klaxon and started the attack run-in from about two miles.

At this I got up on my lookout position again and looked over Cruickshank's right shoulder at a textbook attack.

The flak was coming up fast and bursting round us at the rate of about two a second. As we got closer the two pairs of 20 mm. cannon on the U-boat's bandstand opened up and at about three or four hundred yards Harbison fired his pair of Vickers K guns.

All this firing was exceedingly one-sided. The U-boat's explosive 37 mm and 20 mm cannon could easily penetrate our alloy hull which was nowhere more

than five millimetres thick but all five of the U-boat's guns were behind ten-millimetre steel against which Paddy's .303 bullets would only have flattened themselves.

The run in from two miles would have taken nearly a minute and I just stood there looking at the U-boat and its flak barrage. It made no attempt to dive.

We passed right over the conning tower in a way that reminded me of the practice attacks we used to do in the previous year on a light vessel in the Bristol Channel. Cruickshank could place a Catalina within a metre of where he wanted it, and, it seemed to me, within a half metre of the height clearance he wanted. I clearly remember looking down into the open bridge of the type VII's conning tower.

I thought this must be a certain kill and hurried back aft to the blisters to watch the destruction of a U-boat. On reaching the blisters I met the two crew there and they were most upset. In reply to my query, 'What's wrong?' they said, 'Look!' and pointed to the wings. There, on each side, were three depth charges. They had hung up!

The disappointment of this happening after such an attack, when not one round of the U-boat's flak had hit us, was indescribable. I returned to my radar set to keep track of the U-boat if the sea fog got any thicker.

The blister men reported that the depth charges had failed to release and the skipper climbed to 250 metres at a distance out of range of the flak while John Dickson checked and rechecked the intervalometer and Paddy Harbison reloaded his guns with fresh pans of ammunition.

Knowing the submarine was well and truly in sight, I stood up again to watch the attack. This time all the flak was bursting much closer to us and I was surprised at how thick it could be. We seemed to be flying into a wall of explosions.

The skipper went straight in without hesitation and again the aircraft passed precisely over the conning tower and I felt us pull up for the breakaway. But seconds before this we had caught a full burst of 37 and 20 mm rounds.

I looked back and saw the midship's compartment full of dense smoke in which I could see licks of flame. There were several fire extinguishers in that area but there was also an extra large one mounted on the forward side of the bulkhead at the rear of the navigation compartment. I grabbed this and entered the midships compartment but the two blistermen seemed to have the fire under control.

This fire resulted from a break in the fuel line from the engineer's panel to the tail unit de-icer. The adjacent bunk and various life jackets, pullovers and so on lying on it had fuelled the flames.

I returned forward to the radar set but one glance showed that it was wrecked. I went through the bulkhead door into the nose compartment where I saw Dickie (John Dickson), the navigator, lying lifeless on the floor and Paddy obviously very badly wounded in the legs.

The nose compartment had suffered a lot of damage and I noticed that the windshield in front of Jack Garnett was shattered. Then I became aware of blood and noticed Jack's left hand was badly cut and that he was flying the aircraft.

I went back aft and got the first aid kit. As I broke the seals I suddenly realised how little first aid training we had had.

I got back between the pilots and fixed Jack's hand. I could see that the skipper's trousers were badly blood-stained and assumed his legs had been hit. I found a pair of large scissors and started to cut away the legs of his trousers. I hadn't got far when he went ashen and collapsed. I managed to stop him falling off his seat and someone came forward and helped me get him down on to the catwalk. It was probably Sergeant Fidler, our third pilot, who was making his first ever operational flight.

We carried the skipper aft through two bulkhead doors and over the after main frame and put him on the port bunk. Fortunately it was undamaged, unlike the starboard one which had suffered badly in the fire. We were very glad our aircraft was not fitted with overload tanks on this trip as they considerably restricted movement through the midship compartment. What Cruickshank must have suffered while being carried through the aircraft is too painful to contemplate.

By now, blood was seeping through his heavy flying clothing and I realised he must have bad chest wounds. He had regained consciousness but made no reference to his terrible injuries. I wiped his face with a dampened bandage and gave him a drink of water. Clearly he had lost a lot of blood and some liquid replacement was essential. I knew it would be at least six hours before we would even reach base and the normal rules about giving water to a patient with possible chest wounds were hardly relevant.

The skipper was obviously in great pain. There were supplies of morphine in the first-aid kit. It was a small toothpaste tube like device with a hypodermic needle built in—a very handy one-shot arrangement. I got one tube out, removed the safety cap and remembered the instruction that we must write the letter 'M'—and if possible, the time—on the patient's forehead in blood. Cruickshank realised what I was preparing to do and emphatically responded 'No! No!' He must have realised that his remaining faculties would be seriously diminished by the morphine. So I made him as comfortable as one

can on a canvas bunk and kept him warm with whatever Irvine jackets, spare pullovers and so on that I could find.

Meanwhile, the rest of the crew had set course for Sullom Voe. Garnett and Fidler, the second and third pilots, seemed to be managing the flying and navigation, the second engineer and the rigger looked after the engine panel. Jack had slipped across to the left seat. The left windshield was reasonably intact and the captain's instrument panel had more comprehensive blind flying instruments than did the co-pilot's. There were two other signallers in the crew so I kept out of their way. They had sent messages reporting our attack on the U-boat and the damage and casualties we had sustained.

I went back to the skipper and was relieved to see that he seemed to be holding up. I gave him another drink. By this time, after all the to-ing and fro-ing around the aircraft, I began to wilt a bit myself. I lay down beside the auxiliary power unit and fell asleep.

I was wakened by someone who said we were over base. We had flown down from a place where the midnight sun shines to one where there is at least a short period of darkness. I spoke to the skipper and told him where we were. He immediately said 'Help me up!' I remonstrated, telling him he must rest and that Garnett and Fidler were handling the plane. But he persisted and I could see how determined he was.

With someone's help I started to carry him forward over that frame, through two bulkhead doors and eventually got him into the co-pilot's seat—a terrible trip for anyone with only a fraction of the wounds Cruickshank had suffered. He immediately took command and decided it was still too dark for a safe landing. We had arrived over base at about 0330 so we circled base, at what we were later told by those watching us, was rather an excessive speed. It transpired that the tubes from the pitot head to the airspeed indicator had been damaged and the earlier concern about our ability to maintain height was due to the inaccuracy of the airspeed indicator.

Morning twilight commenced and Jack Garnett, with Cruickshank's help, landed at 0405. Ten minutes later the sun rose.

As soon as we landed we were surrounded by all sorts of marine craft. Everyone was there to help us— the Station Commander, Squadron Commander, Chief Medical Officer and all their staffs. Our aircraft had large holes in the bottom of the bow compartment and under the starboard bunk. These had been stuffed with engine covers, life jackets and so on but we were taking in plenty of water. Our base had a designated beach area at the end of the landing run for situations such as ours and

Catalina flying boat.

Jack Garnett, with a final burst of power from the engines, beached our sinking flying boat.

After two weeks in hospital I returned to Sollum Voe and prepared for sick leave. Paddy Harbison's wounded legs took a month to heal. The skipper was still in Lerwick hospital seven weeks later when it was announced in the London Gazette that he had been awarded the Victoria Cross. His wounds had been terrible and only the brilliant surgery of the RAMC major had saved him.

A total of seventy-two pieces of shrapnel were removed from his chest and legs—mostly from 37mm ammunition. In the six hours return flight to base he must have lost incredible quantities of blood, yet, after all that he insisted on commanding his aircraft for the landing.

John Cruickshank was Coastal Command's third winner of the Victoria Cross.

ICELAND

The distance between the North West tip of Scotland and the coast of Iceland is 330 miles and the Denmark Strait between Iceland and Greenland, though narrowed by ice at certain times of the year, is 180 miles wide.

Although the Germans had occupied Denmark at the same time as they seized Norway they made no attempt to seize Iceland and the Faroes, which were dependencies of Denmark.

Iceland became an Allied base and played a valuable part in the protection of convoys. Roy Baxter recalls:
I served in Iceland for the last six months of the War.

The menace of the U-boats in the Bay of Biscay had been eliminated. The U-boats now concentrated in the waters between the British Isles and Iceland.

Our patrol area was between Ireland and Iceland. We didn't have any sightings up there but that wasn't surprising because most boats submerged for this part of their journey and, anyway, there were fewer boats at sea.

When Germany surrendered it was felt by the powers that be that some of the boats may not have received the message telling them to surrender. In any case it was felt that some boats would not obey the instruction.

Our instructions were:

If we sighted a U-boat on the surface and it surrendered immediately—OK. Otherwise sink it!

SHIP BUSTING

Taking the offensive to the enemy involved attacking his convoys. This evolved over a period and became more effective as time progressed.

At the beginning of 1941 most daylight attacks on enemy shipping were carried out by Blenheims. Beauforts were also used but they had a bad reputation as being troublesome aircraft. Only careful nursing and cosseting allowed the Beaufort to remain operational.

Swordfish were used by the Fleet Air Arm and made some very successful forays.

The Beaufighter seems to have gained the best reputation. Pilots were very happy with its performance.

Jim McSharry tells of his experiences with Beauforts:
The Beaufort was used for a variety of purposes. I did anti-submarine patrols in one in England in early 1942 (January to end of May when we went to the Middle East). We did some bomb attacks on enemy ships and ports. We attacked Lorient with about six aircraft.

A lot of our time was spent patrolling where it was thought enemy supply ships might be trying to slip through and in generally keeping a watch on what was happening. Really, we were pretty raw and we got soft targets to get us used to operational flying.

The crew comprised a pilot, a navigator/bomb aimer, a wireless operator and a gunner. The wireless operator and gunner were interchangeable.

The aircraft could carry a torpedo, although I didn't carry one while in England. We had been taught how to aim and drop them but we were inexperienced crews and torpedoes were very expensive pieces of equipment.

The torpedo was dropped by the pilot who had a sight mounted in the cockpit. The aircraft had to be flown below 100 feet with the nose up at an angle of 2½ deg. to within 600-1000 yards from the target. When dropped it went under and down fairly deep and then came up and ran under its own power.

The Beaufort had been originally designed to be powered by the Hercules engines. These gave something in excess of 1600 hp. However, by the time the aircraft were being built there were more important uses for these engines so the Beaufort was equipped with the Taurus. This was a sleeve-valve engine like the Hercules but gave only 1050 hp. It was used widely for Fleet Air Arm aircraft.

The Beaufort was a heavy aircraft as it was heavily armoured and was meant to carry a heavy torpedo or bomb load. It would not fly on the power of one Taurus. If one engine was put out of action you just had to come down—and in a hurry.

It was claimed that on one engine it had the gliding angle of a brick. A lot were lost over the sea.

Jack Davenport had considerable flying experience in Beaufighters engaged in shipbusting and relates his experience:

• **Rocket Development and Technique:**
We were converted to Beaufighters carrying rockets on anti-shipping runs. The rockets were hung under the wings. During the development stage there was a problem with hang-ups which could mean curtains for the crew.

I did some of the low-level flying for the development of the rocket. In particular I experimented with the method of attack. I had been given a ship, a three or four thousand tonner, painted all over with black and white squares, and moored off the north coast of Norfolk.

I would attack it with various types of rockets. I would sink it and when the tide went out they would patch it up and I would attack it again.

A solid head 25-pound rocket, of which we carried eight, would go through the side of a ship, through a cargo like coal, and come out the other side. It would certainly go right through a submarine. The hole it made when it came out was about a metre across.

There were rockets with explosive heads. All the original thinking was in favour of them. But the work we did proved that, on shipping, you were much better off with a solid head rocket.

When we attacked we did so as a squadron or a section of a squadron. It was a highly coordinated affair. Rarely did we attack individual ships. We normally attacked a convoy of five or six merchant ships with escorts—normally converted tugs equipped with small turrets containing anti-aircraft guns.

In a torpedo attack the torpedo was normally dropped at a range of about 800 yards from a controlled dive. One then had to pull up hard to miss the mast of the ship. Some of our people didn't always manage that.

I took home two sections of mast myself. One was from a very heavy attack in a harbour where there were some large ships and a whole lot of medium sized merchant vessels—about one hundred in all.

We went in and then came out and attacked them going out to sea. The bit of mast was in the centre of the nose in both cases. It ended up against the rudder pedal and wedged there.

Beaufighters were a very very tough fighter. A lot of people walked away from them after some pretty nasty incidents.

• **The Actual Attack:**
The problem was that the escort vessels would be around the convoy at just about the range at which you fired your rockets. That made life uncomfortable because at that point you had to be flying accurately, not skidding from side to side, if you were to hit what you were aiming at.

We had eight rockets and four 20 mm cannon. The technique was to start firing the cannons at about 1000 yards. If the shots fell just short of the ship you kept flying on until the splashes showed the shells were touching the waterline of the ship. Then you pressed the rocket button. You couldn't miss.

This was reasonably easy to do but you had distractions. The Germans developed a technique of firing, from the merchant ships, a rocket which went vertically from the deck. It had an explosive end and was supposed to catch on a wing and explode. They were frightening and were additional to the normal flak.

But our attacks were very effective.

We carried a crew of two—a pilot and a rear gunner/navigator/wireless operator. He did the lot. It must have been hell for him; he couldn't see what was going on. He was more useful looking out the back where any fighters would come from. They were our big worry—they would come in as we were attacking.

SORTIES TO NORWAY

The port of Narvik lay on the Northern coast of Norway. There was a lot of traffic from it—much iron ore was shipped along the coast to feed the German munitions industries. Ships sailing along the west coast of Norway could shelter behind islands for much of the way but there were stretches of open sea where this protection was not available. Moreover, not all movement could take place at night. To do so would involve a lay-up each day and cause unacceptable delays.

Jack Davenport continues:
Flying from Scotland, Norway was one of our hunting grounds. All trips were in daylight.

On that coast, however, near Stavanger, there were three aerodromes on top of the cliffs. The ships would stay in close to the cliffs so the Germans could scramble their aircraft when they picked us up on radar and be ready to meet us.

They were most unpleasant journeys—quite unpleasant. These three dromes seemed to be rest centres for the top German crews. They would be sent to Norway for a month's rest from attacking the daylight bombers over France and Germany. Here they had us—at low level so they didn't have to climb to high altitudes and with only two .303's in the back for our protection.

Jim Savage has a comment:
We were always looking for U-boats, but on one occasion we were looking for a flak ship off the coast of Norway.

Flak ships were old destroyers with the superstructure removed and stacked with anti-aircraft guns.

There was some flak flying round so we felt we'd found it and returned to base. We think we were credited with finding it but the crew—except the skipper—were not always advised of the results of the patrol. It was left to the skipper to tell us if he was permitted. Only the skipper, the navigator and the first wireless operator went to briefing.

ATTACKS ON SHIPS OF WAR

Besides attacking convoys and lines of supply, the planes of Coastal Command were used, from time to time, to attack ships of war.

After the D-day landings there was considerable concern about German naval activity against the ships supplying the beachhead. One of the most effective German weapons was the E-boat.

These small but lethal craft had used their torpedoes to harry shipping on the east coast of England throughout the war and were in an ideal position to wreak havoc on the invasion fleet. On 6th June—D-day, they sank the Norwegian destroyer **Sford.**

Jack Davenport was engaged in an action to curb this menace:
Shortly after D-day the shipping round Mulberry Harbour[1] was suffering from E-boat attacks. These German craft were much faster than the British boats. In fact one squadron of E-boats used this to advantage by sitting off the harbour to taunt our boats into attacking them. At this, they would evade the attack and slip past the attackers to torpedo our ships in the harbour.

This got past a joke so Command asked us to work out some counter action.

We devised a technique of formation flying at night. This may sound ridiculous but it worked. We formated on the glow of the aircraft exhausts. The aircraft had to be very very close.

Because we would have to dive to low level at night, we also measured the reading difference between our radio altimeters and the normal barometric altimeters when diving. With all this worked out, we arranged for a radar-equipped Wellington to drop flares behind the E-boats so they were readily visible.

Our Beaufighters carried out two of these attacks. Both were highly successful and the E-boats were scattered.

1. One of the two artificial harbours constructed on the invasion beaches. They were made from gigantic floating concrete structures. These were towed across the Channel, carefully positioned and sunk to allow the rapid unloading of ships at the invasion beaches.

AIR SEA RESCUE

Coastal Command was often asked to provide resources for the rescue of aircrew from the sea. Flying boats of the Command were at times required to make landings on the sea to rescue downed crews.
David Corthorn trained for this activity and talks about the equipment available:

We trained to carry out Air Sea Rescue operations on Warwicks. These were very similar to a Wellington—geodetic construction of the fuselage which was covered with a skin of fabric. There were four in the crew—pilot, navigator, wireless operator and rear gunner.

We carried an airborne lifeboat and could carry, if necessary, a 2000-pound bomb load.

The lifeboat was about forty feet long. It had twin engines and was dropped from a height of about 1000 feet suspended from six to eight parachutes. When it hit the water an electric slip was actuated to free the parachutes and an internal battery connected to allow the engines to be started. At the same time four lines, with floats at intervals along their length, were fired, one each from bow, stern and each side. These were for drifting or swimming crew to grasp and pull themselves to the boat. Each float on the lines had an arrow painted on it pointing towards the boat to guide the victim in the right direction.

Because of the very low temperatures in the seas around England, one could not survive long periods in the water so the airborne lifeboat was a useful device to get men out of the water and give them a means of making their way towards safety.

Besides the engines, the boat was fitted with a mast, mainsail and a jib. It came complete with a book of instructions on how to sail it as well as supplies of food and drinking water.

PIGEON COMMAND

Not only humans, but also animals served in the Allied forces. Pigeons played a useful role, as related by John Appleton:

Each aircraft of our squadron operating over the sea carried two pigeons. As they were for communications they were the responsibility of the signallers.

Each was kept in a bright orange metal container somewhat larger than a shoebox. The container had a lid which, when fitted, made the box watertight so the pigeons could survive ditching in the sea.

The pigeons were trained to fly home to our base. In the event of a forced landing in the water the pigeons could be released with a message giving the position of the landing. Each pigeon had a blue plastic container to contain the message attached to a leg. The message was written on what was surely the smallest form used by any of the armed services—Form 1326. It was the size of a bus ticket and had a space to write the date, time, aircraft call sign and whatever information was to be sent.

The first pigeon was to be released at dawn, as soon as possible after a downed crew had got into a raft and had collected its thoughts. Any position given might not be accurate but at least would give base some information. This would be particularly important if there had not been an opportunity to transmit any radio signals before the landing.

MALTA

Control of Malta was vital to maintain the line of sea supply through the central Mediterranean but it was not possible to retain a fleet there because of the menace of overwhelming air attack.

Nevertheless Malta was maintained as a base for submarine, naval and air attack against enemy troop movements in the Mediterranean. It was particularly necessary to the harassment of enemy shipping seeking to reinforce General Rommel's forces in Cyrenaica.

The island was subjected to intense air attacks which increased in severity when, early in 1941, the Luftwaffe started to attack the convoys bringing in vital supplies. Malta was awarded the George Cross in 1942 as a tribute to its valiant resistance.

Jim Savage takes up the story:

I flew to Malta twice in 1942. The first trip was taking Richard Casey there. He was on his way to Egypt as Minister of State in Cairo—sent out by Churchill. He also had to go to Malta to discuss whether General Dobbie should be relieved.

We flew without escort and followed the Spanish coast as far as Cape de Gata and then across the Med to Cape Bon. We flew at night and the lighthouse at Cape Bon was a wonderful navigation aid.

At Cape Bon we changed course and headed up between Pantelleria and Sicily. We had to go down to water level there because of the radar in both places. They had Me-109s and Me-110s there and we didn't want to meet them.

We skimmed along until we came close to Malta and then had to give a prearranged signal with our landing lights—flashing them along the south east coast of Malta. We landed at night on that first trip and sent Casey ashore. There were air raids while we were there.

We refuelled in good time and flew back to Gibraltar—no hazards this time.

A few days later we took Lord Gort aboard to fly him to Malta to take over from General Dobbie. We had damaged the port float on the first flight and had it repaired by the RAF at Gibraltar. After we were airborne I noticed one of the braces [wires] was stripped again. I went up to the skipper and whispered in his ear, 'Have a look at that bloody port float.'

Lord Gort was standing behind him and the skipper told him, 'I think I'll have to go back.' Gort said, 'Mr Stokes, I am completely in your hands. Whatever you have to do, do it.'

So we turned back and jettisoned a large part of our fuel because we had just taken off. The fuel went out in the slipstream—I was thinking of how I could use it in a car in England and here it was just pouring out. As we landed I told the skipper to be very easy on the float and he was.

The RAF worked all night on the float—largely under my supervision.

We did a test flight next day and everything was all right and we took off again for Malta around 1 p.m.—we had to reach there by night.

On the way I was in the wardroom with Tommy Stokes and Lord Gort. He patted his briefcase and said, 'Mr Stokes, you will be very pleased to know that I'm taking the George Cross to Malta 'and added, 'I have it here in my briefcase.' Neither of us saw it however.

On our approach to Malta we had to circle a small island at 500 feet to be recognised before they'd let us land. It was a very poor flarepath, just three vessels with a small light on them which went out virtually as we landed. There was oil all over the water—a tanker had been sunk there a day before.

We got Lord Gort ashore with all his party and the skipper went with him. There was an air raid on and Tom Stokes had his tin hat jammed down on his head all the way. Lord Gort was sworn in within the hangar.

We had lost two aircraft at Malta. One had been sunk in the bay by Me-109s in daylight. The other was an RAF aircraft being ferried out to the Middle East by one of our crews. It was out on the tarmac (slipway) when a Stuka dropped a bomb right behind it and blew the tail off. It should have been taken into the hangar but the RAF people there were a bit slow and groggy from the pounding they had had from the Germans and Italians and they left it at the end of the runway (sealane).

We started refuelling but the refueller had been sunk and we had to refuel from 44-gallon drums with an air raid well and truly on. We were pumping so slowly with the pumps that we would have been there next day.

We didn't want to be there at daylight and lose our plane. Monty Blue got the idea that we could suck up the fuel with the auxiliary power unit—which was in the mainframe—and was used to suck the water up out of our bilges. So we ran a hose down into the first 44-gallon drum and 'swoosh' and up went forty-four gallons.

It still took far longer to refuel than it should but we eventually made it. We took aboard the passengers—the Governor, his wife and daughter, refugees, wounded and otherwise and were heavily laden.

We taxied out and started our run but couldn't lift off. The oil was holding us to the water. It was still—no breeze, nothing.

We went right out of the heads into the sea and we still couldn't get her up. We were worried about the repaired float and I was worried about the planing problem (i.e., the pressure on the underpart of the hull). I conveyed my concern to Tom who was also worried as the the cylinder head temperatures had gone up.

He said, 'I'll give it another run.' It took us four and a half minutes to get off—the longest it ever took. As we were airborne an aircraft came swinging towards us—an Me-109 or 110—and Tom Stokes made for a low cloud. We got into it and followed it round the island at sea level.

General Dobbie was a very religious old man—a member of the Plymouth Brethren. I set him down in the wardroom and left him there with his Bible on his knees. Every time I went through the wardroom, to the nose or some other turret, he would say, 'Are we through the danger area?' I'd say, 'No, we've still a way to go.' The last time I went through I said, 'We've just passed Cape Bon.' He closed his Bible and said, 'God be praised.' and went to sleep.

We had a good rest when we got to Gibraltar.

In early 1942 there was great concern that Malta might not survive without additional air resources. The aircraft carrier **Eagle** *managed to fly thirty-four Spitfires into Malta during March 1942. More were urgently needed.*

Churchill persuaded Roosevelt in early April to make the American aircraft carrier **Wasp** *available. Jim Savage, who was detached to Gibraltar at the time, takes up the story:*

We convoyed the carriers taking the Spitfires into Malta. We left them at Cap de Gata—we were a safeguard against submarines. There were no enemy fighters in our area of escort.

There was the American aircraft carrier *Wasp* and the British aircraft carrier *Argus*—each with its flight deck covered with Spitfires. There were destroyers and the battleship *Renown*. I can remember being taken over her when she visited Sydney well before the war.

Jim Savage (front row, right) and some of his crew and ground staff at Gibraltar.

One of the greatest sights I had was when the crew of the *Wasp* came out and waved to us as we left them in the Med.

The delivery of the Spitfires virtually broke the siege of Malta. They flew off the carriers, refuelled quickly and were in the air when the Luftwaffe next came over. The Germans had one hell of a shock. Up to then the Spits had been brought in packing cases and had been shot up before they got off the deck.

126 Spitfires reached Malta during April and May.

COASTAL COMMAND —ANOTHER SIDE

All work and no play makes Jack a dull boy. When the boys of Coastal Command relaxed they made sure there was plenty of play and their company certainly was not dull.

Jim Savage tells of some exploits. He chose to call the first story 'A rose by any other name':

Nicknames—nearly everyone had one. Some were quite affectionate, others scatological, a lot most descriptive, but dull ones never—well, hardly ever. Quite a few had good stories behind them; for instance, how would anyone end up with 'Screaming Skull'?

Mine didn't reach me until mid-1943—well after I had joined the squadron in 1940 and it was simply 'Tombstone.' Now there are many versions of how I got it but the truth is that I was at a loose end one evening in July with my aircraft on deck for inspection. One of my still surviving mates from my original crowd poked his head round my door and said, 'What about a drink?'

Being cursed with a weak will in such matters I answered that it was a great idea. As it was a perfect evening we decided to have a walk to the little pub down at Hooe. There, in a happy and convivial atmosphere we took our ease and probably more than a few good British beers. As it was Double British Summer Time it was still quite light when the landlord called 'Time gentlemen please.'

We finished off our last drink unhurriedly for Reg and I were at the sentimental stage by then and that last drink had been dedicated to good mates who were no longer around. We didn't want to cut it short. However, the landlord did that for us and we found ourselves outside and heading for home.

The last bit I remember was the two of us feeling very sorry for ourselves with Reg declaring he didn't have a chance of getting home to Australia and me saying I didn't think I would either. We were, at that stage, completely in agreement with one another in not wanting to end up as coral in the Bay of Biscay.

At this point we were level with a monumental mason's establishment on the left-hand side of the road as you left Hooe village. The last words I remember uttering were that, even if I did end up as coral, I would not be without a tombstone.

I woke up the next morning with a splitting head and only a hazy recollection of the night before—and absolutely nothing of the journey home after leaving Hooe, up the hill, past the security at the station gate and into my bed.

I didn't see Reg because his crew went off for three days bombing practice at Stranraer up in Wigtownshire in Scotland. On the fourth day, or thereabouts, he came to me and invited me to read a small notice in the local paper. It offered a reward for the return of a tombstone to the mason at Hooe and any information on its disappearance. I asked him what that had to do with me.

He said I surely must remember picking it up, carrying it up the hill and tossing it over the wall of the churchyard at the top. I told him he was nuts but agreed to go with him and check it out. Off we went to the church and into the little cemetery behind the stone wall. It didn't take long to find what we were looking for in the long grass. When I saw it everything came back as clear as day.

We took off for the Hooe pub to think it out. We stayed until closing time but drinking slowly. As darkness fell we returned to the churchyard. You know I could hardly lift that damned tombstone let alone carry it. It took the two of us all our time to get it downhill in the blackout and put it back in the mason's display.

It is a sad thing when you can't trust even your best mates. That damn fool Reg couldn't keep his mouth shut and the story went through our crowd like a bushfire and that's how I got lumbered with the name 'Tombstone.'

If when we march on Anzac Day some silly ass yells it out as we go past, just ignore it and don't look in my direction as I am quite a respectable bloke these days.

Jim has another story:

'Be prepared to be away a week.' said the Flight Commander and, no sooner said than done, the crew were on Plymouth North Road station waiting for the London train en route to Rochester in Kent to pick up a brand new Short Sunderland. We had relinquished E for Eddie, one of the squadron's original boats, and seen it off to Northern Ireland to be scrapped. We were looking forward to a 'factory fresh' aeroplane.

There had to be an overnight stay in London and even in early 1942 and despite the Blitzes much fun could be had. When we caught the afternoon train for Rochester we were, with the exception of the skipper who was a very moderate bloke, just a little the worse for wear. We had recovered somewhat by the time we had arrived at our destination and settled down quite happily in our billet, a very old and famous pub. The skipper and our three other officers were set up just down the street from us in rather more pukka digs but, in fact, rather preferred the less stuffy atmosphere of the crew's pub.

When our skipper and first pilot presented themselves at the Short works early next morning they were told of a hold up in delivery. How long? About three or four days.

We received the sad news quite philosophically and prepared ourselves as best we could to carry out the skipper's wishes that we enjoy ourselves. He simply wanted the first fitter to report to him each day. He finished up by asking how we were placed for money.

Whenever we were stuck in some outlandish place due to bad weather or some other flying restriction this was his usual question. We, being RAAF, had no way of being paid so we were invariably short. He, on the other hand, appeared to have access to bank funds anywhere—even as far afield as Gibraltar, Malta or Alexandria—and we very soon had whatever we asked for. No strings, no interest, just pay him back some time after we returned to base.

All this was in marked contrast to many of the captains—particularly the old Point Cookers—who didn't take much interest in their crews and certainly didn't concern themselves with their on-shore problems. No wonder we reckoned our skipper was great.

Now the decent thing to do would be to draw a veil over the fun and games that followed. They were not restricted to Rochester but extended to Maidstone and a few other spots around. They centred mainly on pubs and girls—or girls and pubs, I'm not quite clear. We had, I must admit, a hell of a time.

But Nemesis was waiting in the wings and early on a Friday morning we were called and told to be ready for transport to the Short factory at 0930 hours. Now I, being the youngest and by far the best behaved (I am prepared to admit this in all fairness) was not so badly off except for a slight headache and a mouth like the bottom of a bird cage. But the others? It was really quite sad.

Thankfully there were no handing-over ceremonies and the two-hour wait before the pinnace took us to the aircraft was a godsend. It only needed the skipper, the first fitter and me to sign for it and it became the property of No 10 Squadron RAAF.

The crew went about getting their stations in order and it was not long before we slipped moorings and I reported to the bridge that 'All hatches were secured.'

Then the skipper called us up on the intercom and broke the news that he would have to have the aircraft 'up on the step' as he came round the bend in the Medway because the take-off distance was restricted by a bridge. He didn't tell us that he couldn't bank at all if he thought he couldn't clear the bridge—the gasworks were to port and Rochester Castle was to starboard.

All this and hangovers too. There was nothing to do but huddle down, hold on and pray. We went roaring round the bend up on the step in a cloud of spray and on full boost, straight for the bridge. My most vivid memory is of the pink upturned faces as we cleared it. I stopped praying—we were airborne.

There didn't seem to be a course set, or at least the crew didn't know about it, and we flew on just under the murk which was at about 1200 feet over a snow-covered landscape only relieved by villages, farmhouses and an occasional copse. The crew were intent on getting the feel of this bright new thirty-ton aircraft and did not take much notice of the passage of time until the skipper came on intercom to our observer, a real old squadron man.

All he wanted to know was where we were—a quite reasonable request. Stan answered that he was buggered if he knew. An answer which, while honest, was no use to the skipper.

After a few pungent words from the bridge it was considered advisable to make a goodly circle in the hope of picking up the 'iron compass'—the Southern Railways route, which was known to run from east to west—

and follow it home to Plymouth.

We found it and a few minutes later the skipper called up the crew and asked if anyone had seen or visited Salisbury Cathedral. We thought that a silly question until he told us it was up ahead and we were going to have a good look at it

He took us down to 500 feet and we flew a tight circle round it getting a truly magnificent view as we went over the famous Constable water meadows, the town and even the ruins of Old Sarum. The skipper had been there as a youngster with his parents and was really bucked at seeing it again—from a different angle too.

I don't know what the local inhabitants thought, but that great lumbering Sunderland at that height must have been some sight. It wasn't every day that thirty tons of aircraft zoomed around their cathedral's spire.

We duly arrived back at base and some rotten spoil-sport had dobbed us in. The skipper and observer got a good bawling out by the flight commander. It could have been worse if our squadron had not been—on account of its serviceability and general performance—in the good books of 19 Group Coastal Command.

So that is how a hungover crew came to see Salisbury Cathedral.

When I went to see it after the war it didn't look quite the same.

Sunderlands in formation.

2

THE MIDDLE EAST & MEDITERRANEAN

Before Italy's entry into the war, the Middle East Air Force had been gradually building up its fighting potential. It had no defined operational role. The build-up was limited and the force still lacked satisfactory fighters and long-range aircraft when Italy entered hostilities.

The area to be defended was immense. When, in May 1940, Sir Arthur Longmore took up duty as head of the Middle East Air Forces, his command embraced control over RAF units in Egypt, Sudan, Palestine, Transjordan, East Africa, Aden, Somaliland and Iraq. He also had units in Cyprus, the Balkans, the Red Sea and the Persian Sea. Maritime operations by aircraft in the Mediterranean Sea were also his responsibility.

The forces at his disposal were pitifully thin: Egypt/ Palestine—40 Gladiators, 70 Blenheims, 24 Bombay and Valentia transports, 24 Lysanders and 10 Sunderlands; Kenya—3 squadrons of the South African Air Force comprised of one of Gladiators, one of Battles and, strangely enough, one of Ju-86s.

Italy entered the war at midnight on 10 June 1940. On 11 June the air war in the Middle East commenced with attacks on Italian bases.

The struggle was to be a long and bloody one. This chapter tells this story, not as a full history of the campaigns, but as a series of eyewitness accounts from the scattered corners of the command.

WAR IN THE WESTERN DESERT

3 September 1939 to 12 August 1942

On 13 September 1940 Mussolini forced his Italian Commander-in-Chief in Italy to move towards the Egypt/Libya border.

At Sidi Barrani the Italian forces dug in and ceased their advance.

The British Commander-in-Chief, Middle East (Sir Archibald Wavell) planned to attack the Italians in December. To support this offensive Wavell sought more air support. Flights were withdrawn from Aden and, although planes had been sent to Greece, it was still possible to re-equip some squadrons with more modern aircraft. This was because, in October 1940, the Takoradi route was opened up to fly aircraft to the Middle East, avoiding the long and dangerous ship journey round the Cape of Good Hope.

Planes could be flown from Takoradi, on the west coast of Africa (in present day Ghana), to the Nile and then north to Cairo. A gradual build-up of air fighting strength was possible.

During the build-up period George Gray served with three different RAF squadrons—Nos 45, 113 and 11. He comments:

There were four Blenheim squadrons and in some of the early pushes I started to work with the then Army Cooperation Unit. We were maids of all work, attacking tanks and trucks and carrying out low level attacks on the roads using our 20 mm cannon and machine guns to strafe whatever came in sight. Incidentally, the cannon was sited almost in the navigator's lap.

George then moved to 55 Squadron RAF which was located in the Western Desert, and, from February/March 1941 was given the tasks of strategic reconnaissance and attacks in close support of the army.

These operations were mounted with Blenheim aircraft—some being used as 'fighter' Blenheims and others mainly for bomber/reconnaissance.

There were no true stations for the air force in the desert. Improvised landing grounds were built as the army

Captured SM.79 (Savoia-Marchetti) Italian bomber.

advanced and retreated. The strips were primitive and so were the living conditions.

He takes up the story from the day of his arrival on the squadron:

Interesting things happened on the day I joined the squadron at LG 15. The latrines were built of sandbags on three sides with the open side facing the aerodrome. I soon discovered that this was a dangerous spot because half a dozen German aircraft came over and strafed the drome. They managed to catch one of our flight commanders on the throne and he couldn't run because his trousers were round his ankles. We suffered some casualties as a result of this raid so retaliation was envisaged.

Four or five aircraft were detailed to fly south to Giarabub, an oasis at the bottom of the rise which separated Libya from Egypt, deep in the desert with a landing strip on a flat stretch between rises in the desert. It was very dangerous for landing and taking off—although that was par for the course in those days.

We went down at midday with a view to a night strike on the road south from Benghazi down to Marble Arch. My flight was the Blenheim fighter wing and, as luck would have it, our aircraft were given, temporarily, to the other flight for bombing duties and none came back in one piece.

So we took off in the reserve aircraft. It was necessary to give the Blenheim maximum boost to get over the hill at the end of the very short runway.

Off we went, only able, with a dicey port motor, to reach a maximum height of about 800 feet above the undulating country. The trip got worse and worse. The motor caught fire but we took fire action and put it out but the propeller was still turning which slowed us down and caused us to lose height.

As darkness came and we were still half an hour out we started to call up on our emergency radio frequency but with no result. Then I reckoned we were hitting the drome—it was just a strip of dirt, no asphalt and no flares. Although we hit it on the nose we suddenly realised, having fired the colours of the day, that there might be intruders about. It turned out that there were 79s [SM.79s—Italian three-engined bombers] bombing the drome and so we attracted some local ack ack, which was understandable.

We were going round again when the motor exploded and fell out. We were probably 200 feet above the drome and it was about 8.30 p.m.

Second Lieutenant Butch Sommersgill did a magnificent job of putting the Blenheim down on a section of wadi with the nose ending up by a large rock. There was petrol all over the place and I remember scrambling out one side and Butch going out the other. The air-gunner was more seriously injured. Twice we went looking for help.

I wrapped the gunner in parachute silk—it was pretty cold and I remember there was a sort of sandstorm that made visibility a problem.

A NZ Army crew, in what we called a Quod car, pulled up, probably one or two hundred yards from us and said, 'Come on you B's! Hands up!' To which we replied in some fairly rich Australian that they should come and get us—which they did and took us and the gunner to a casualty clearing station in the Badoush box.

The army had set this up as a stronghold, to operate in defensive mode if necessary, but which would provide a post for attack when they were ready to go. The CCS [casualty clearing station] was underground and I can remember one part of my stay there.

They had put me on a stretcher and someone said 'Where are his dog tags?' I pulled them out and read my name off and a face came over the top of me and a voice said, 'Georgie boy. Are you from Drummoyne?' To which I replied, 'No. I'm from Sydney, but I played for Drummoyne.' To which he replied, 'You're the man! You and your club entertained me at your football ball when I was visiting with the 1937 Springboks.' And then he said, 'I'm Louis Dubroux and a centre three quarter from Stellenbock University.' That's the last I saw of him.

That meeting was the one plus of the crash for me because I had bad results from all my cuts, bruises and scratches. I was damaged all over because the skipper had landed on the belly on one engine in the dark and things had broken up all round us.

I was sent from the CCS to hospital. I spent some time in the hospital train going back to the Delta area where I was treated in two different British hospitals, one at the British General Hospital in Cairo and the other out near the aerodrome.

As I was being carried, on stretchers, up and down during air raids, from whatever floor we were on to the underground shelter, I ran into an old mate from Drummoyne who had been in Tobruk. He was going up the stairs as we were coming down. It was Tom Pulbrook who said, 'Why don't you get into an Aussie hospital?' I promptly requested this privilege and I was transferred to 2 AGH outside Cairo, on the canal.

Being the only air force bloke there, I was treated pretty well. I was in a ward with Tobruk and Syrian casualties and they were all very severe. I was the healthiest bloke there. But then I developed pretty severe blood poisoning in my legs which were in plaster. So I had a few months in hospital while that came good.
I spent two or three months out of action. I then had a medical examination at Heliopolis. I went in with a little RAF fighter pilot who came in with a pretty bad limp. We went through a full medical and when he came out I said, 'How are you going?' He said, 'All right! How are you?' I replied the same way.

We were both reported fit for duty immediately. There was a shortage of aircrew at that time.

I went back to 55 Squadron, still on Blenheims. It was now a GR [general reconnaissance] squadron and I joined a new crew. In the meantime my old crew had been lost. They had been killed in action with another navigator in my place.

Ted Eagleton's introduction to the Middle East was unusual:
We arrived in England on Christmas Day 1941—stewed prunes for Christmas lunch. We were transferred to Bournemouth and, while there, it was an extremely cold winter. We were awaiting postings and they called for volunteers for the Middle East.

We went out because they said they wanted fifty aircrews. As it turned out, they wanted fifty airscrews! But we'd arrived.

We were trained on twins and weren't allowed to fly on singles. Rommel was coming through so we were sent to a place called Gordon's Tree at Khartoum, in the Sudan. It was quite an adventure going up the Nile in paddle-steamers.

When we got to Cairo Japan had entered the war and we wanted to get back to Australia to fly. Lord Casey was in Cairo at the time and we got to see him. He got us a posting in forty-eight hours—back to England. As a result we went round by the Cape. Another Christmas aboard ship!

13 August 1942 to 12 May 1943

• The Battles of Alam Halfa and El Alamein

General Montgomery assumed command of the 8th Army in the Western Desert on 13 August 1942. He writes that his orders from General Alexander were quite simple—destroy the German forces under Rommel. This he proceeded to do.

Montgomery's first major confrontation with Rommel was at Alam Halfa.

Rommel's goal was to capture the city of Alexandria. The ultimate goal was to drive the British out of Egypt in the short term and the Middle East in the long. The capture of Alexandria would be a major stepping stone since it would remove the British fleet from the Mediterranean Sea and make the defence of Egypt immeasurably harder. To succeed he had to defeat the 8th Army which, in August 1942, had two defence lines— one at El Alamein, the other on the Nile. A weak spot must be found and attacked. Rommel selected a point between the left of the NZ Division at Alam Nayil and the edge of the Qattara Depression.

He attacked on 30/31 August. Rommel's aim was to bypass the Alam Halfa Ridge and sweep north to cut off the Eighth Army. Montgomery strongly manned the ridge and heavily mined the ground to the south so that the German forces would be stalled with strong forces on their flank.

The British commander was determined his tanks would not rush out and be mauled by the German armour. He ordered his troops to stand firm where they were— neither advance nor retreat.

The battle raged from 31 August to the 6 September. The intervention of the Desert Air Force was, according to Montgomery, 'a most important factor which forced his [Rommel's] withdrawal.'

The Battle of Alamein commenced on 23 October 1942 with an opening barrage of nearly 1000 guns. It finished at dawn on 4 November when Rommel disengaged and retreated. Rommel gave, as one of the reasons for his defeat, the British air supremacy which 'created severe limitations on the use of mechanised forces.' The British, he wrote, used 'continuous air attacks by powerful waves of bombers.'

At the time of Alamein there was a total of 96 Allied squadrons in the Middle East. The force was comprised as follows: 60 British [including the Fleet Air Arm], 13 United States, 13 South African, 5 Australian, 2 Greek, 1 Rhodesian, 1 French, 1 Yugoslavian. George Gray tells of a raid in the Alamein battles:

Things were pretty bad in those days. We helped with the first holding battle at Alamein with close support bombing. The final stages of the conversion of 55 Squadron were such that, for two or three months before the main battle of Alamein, we were based at LG 86, west of the Cairo/Alexandria road and midway between the two cities. This was no more than thirty miles from the main bomb line where the battle was. That line ran from Tel-el-Eisa to the Qattara Depression in the south, where the Great Desert prevented anyone getting round—except the rather intrepid Rhodesians who used to go sweeping through it.

Now we were involved in containing the strike that Rommel made with his 15th and 21st Panzer Divisions, the 90th Light Infantry Division and a very large number of AA guns (all 88 mm). The Germans drove through a weak link in the Alamein line. Montgomery said he'd let them through—I'm not so sure—and they were driving for Alam Halfa—a ridge where XXXth Corps had their headquarters. The cooks and drivers and so on got into the trenches on top of the hill and joined in the battle.

We were turned loose on equipment and guns right out in the open with no camouflage. There were two RAF Baltimore squadrons, one South African squadron and four Boston squadrons engaged. We usually operated, at that time, in boxes of six, or two boxes of six in a larger box of twelve. Sometimes we used formations of twenty-seven—three vics[1] of nine. Straight and level flying, close formation. The tactic was very effective because, as you ran into your bombing run, you spread 100 yards from your leader and the second drop started dropping where the leaders had ended so that you got a square of bombed ground. These explosions were devastating for trucks, personnel, and even light tanks.

The drop is recorded in Montgomery's story. 55 Squadron, in ten days, did 352 actual sorties [sortie was one plane going on a trip]. Sometimes three boxes of twelve would go in a day. We were copping a lot of flak with many aircraft damaged. Ground staff were getting them back in the air in a matter of hours. We were taking fighter escort, a top cover of Spitfires or Kittyhawks (mainly Kittyhawks), a medium cover of Kittyhawks and Tomahawks and a close cover of Tomahawks mainly loaded with bombs. As we came away they went in and dropped their load individually, came out and joined us on the way back.

The troops were sometimes as close as 800 yards from our aiming points and, knowing the difficulty of finding landmarks in the desert, the Army's Long Range Desert Group would go out and put down half a dozen petrol cans in the open in the form of a letter like L or A and we used that as a run-in point for a target such as camouflaged gun emplacements. We had some spectacular results on those.

Then we started to operate also on dumps behind the line and 55 Squadron was detailed to go for a dump round El Daba at 1510 hours on 6 October 1942. The force consisted of a box of six from 55 Squadron and a box of six from another Baltimore squadron. Our job was to knock off fuel dumps covered in sand. We knew where they were—humps in the sand and supplies that had been a base for the British Army during the earlier back and forward flow—the Benghazi Handicaps.

It was rumoured also that there were one million bottles of Australian beer buried beneath the sand because NAAFI [the canteen service] had set up a way station there as it was a comfort stop on the main coast road.

We went out over the front line, which was contrary to standard practice. We went over at 7000 feet, straight and level, in broad daylight. We got over Alamein station and then to what became known as Tel-el-Eisa when all hell broke loose.

We had been very upset about flying straight and level. We were number two. We lost three aircraft out of six at that point and reformed with our crew taking the lead and despatching our bombs. Then we had to scramble back through the flak at that height over the line.

One plane exploded alongside Alamein station. Another crew, with pilot Denis Heathers, got downed in no-man's-land. He stood by his aircraft and the Long Range Desert Group went out at 40 mph and picked him up. They were being shelled all the time and how they got Denis back to the squadron I don't know.

Back at base we counted seventy or eighty holes in most of the aircraft which got back.

1. A vic was a formation where the aircraft flew in a 'V' group with the point of the 'V' in front and an aircraft on either side, at the end of each leg.

I was parachute officer and had to go down and organise things. There was a bit of a commotion as one of Denis Heather's crew had come in and asked for his chute to be repacked. He wanted to send it home to his girl-friend. He understood that, if you escaped using a parachute, it was yours to dispose of as you wished. I had to convince him that that was not the case.

This raid had been an exercise in stupidity in leadership. Thereafter we flew at heights which were much more reasonable and weaved. This gave us much more chance to evade the very very skilful and very very accurate and well-directed AA fire. The German gunners used to hang off until they were sure they were going to hit you and then hit you. They were magnificent gunners, the 15th, 21st and 90th Light AA.

• Supply Problems after the Battle of Alamein

One of the problems of the breakthrough after the victory at Alamein was that of supply.

Transport Units played an important role in meeting the problem. Flying from makeshift airfields, the pilots encountered problems peculiar to desert warfare, some of which are described in the following pages.

Warrant Officer (later Flying Officer) Jack Stronach was, at this time, attached to an RAF squadron—No 117—and flew Hudson aircraft. He tells how he flew water and ammunition into Tobruk (El Adem) airfield and of the hazards of these operations. His squadron was quite often the first to land at these advanced strips.

The first problem he met was with the transport of water:

On 25 November 1942, while I was based south of Tobruk at El Adem, I was detailed to fly a short distance to another aerodrome. Before I left, the aircraft was loaded with empty German water cans. There was a large number as they did not weigh much. On arrival they were all filled up and loaded on the aircraft. Water weighs 10 lb per gallon and I knew I had a heavy load. But it was only when I started the take-off run that I got the feeling we were overloaded.

The aerodrome boundaries were marked with 44-gallon drums. I put her right on the edge, ran her up to full throttle on the brakes and then let her go.

I was still on the ground when I went past the other boundary. Jesus! You know I was overloaded all right! The gullies—they called them wadis over there—started coming close. I had run out of runway so pulled back hard on the control column and dragged her into the air. It was a close shave! The strips were just too small for Hudsons carrying heavy loads.

Another problem which Jack faced was the delivery of ammunition. This task nearly ended his operations:

On 29 November 1942 I went on a trip to Tobruk carrying bombs as cargo. I was in formation of other cargo aircraft and a fighter escort. We were flying right down on the deck to get under the enemy radar.

We were about seventy miles from Soluck, which is just south of Benghazi, when the navigator tapped me on the right shoulder and pointed to the port engine. Oil was pouring out. It wasn't much longer before he touched me again. It was billowing smoke. He touched me again: 'She's on fire!' So I had to do something or I'd take someone with me. I had to break formation and try to gain some height.

I looked for power in the other engine. You wouldn't believe it—it started to conk on me too. So I was sitting with 12 x 250-pound bombs behind me and no real power to keep me airborne.

So, down with the undercart and hope for the best. Luck was with me again. I put the wheels on the ground and was running along beautifully. I had the speed down to round 40 mph when I hit the undercarriage on a large boulder which took off the port wheel.

Dust was all around us. The navigator thought he was in heaven and I had to shake him to bring him around. The heat was terrific.

We got out the back door and I said 'Run!' and we'd gone about a hundred yards and got down behind a pile of rocks and 'Whoom!' went the tank. She was really going up now and it wasn't much longer after that when the bombs went off. We were very lucky. We were seen by a South African patrol who had a vehicle which picked us up and took us to Soluck.

Jack Stronach and his bombs.

• Maintaining Aircraft Strength

Replacement of aircraft lost in combat or accident relied on the provision of planes from overseas—either Britain or America.

Ferry command was entrusted with the flying in of these replacements to the Middle East. Gerry Judd and Jim McSharry took part in these operations.

Gerry Judd relates his ferry experiences:
After training at Charter Hall, near the Scottish border, I went on to Ferry Command flying Beaufighters out from Cornwall. You did a flying time test to ascertain endurance as every aircraft had a different fuel consumption. Some, on a full tank, would fly for four hours, others would fly for five. You flew over the Irish Sea until almost empty and then headed for home with very little in the tank.

When you had established that you could make the distance you set out on the trip across north west Africa. You'd fly south across the Atlantic and turn south west to make a landfall which was, hopefully, the southern most point of Spain—Cape St Vincent. Once you got a visual on that you headed for a drome near Casablanca.

You'd start out flying south in jacket and flying boots. Now it was blazing sunshine and glare so you'd change into light shorts.

Your aircraft would be refuelled and then you'd fly from Oran to Algiers and then on to Castel Benito and Cairo West.

There you got a Dakota and flew back the same way to England. This was a hell of a long trip. It was bitterly cold. I remember on one trip a Canadian friend of mine said he couldn't stand it any longer. We were in a Dakota, which had no heating, and it was extremely cold. We had our own parachutes so we pulled the ripcord on his and let it billow out, then wrapped ourselves up in it to keep warm. He said, 'If we have to bale out mate, I'll be hanging around your neck.'

Jim McSharry continues the story of ferry operations:
I went, initially, to the Middle East. There was a great need to ferry aircraft to and about that area. At Takoradi, near Lagos, the aircraft were assembled. They came out partly assembled as deck cargo. The wings were put on at Takoradi where there was also a test unit.

The ferry pilots would pick up the aircraft and fly them across the middle of Africa, then down the Nile Valley to Egypt where they were readied for action. The Takoradi run was the main job for my first unit. I did two trips on that run.

I was next in a unit whose job was to fly good new aircraft up to the front-line squadrons in North Africa and bring shot-up ones back for servicing in the Middle East. I must say that the second part of the operation was the more dicey. But the secret of being a successful ferry pilot was to be adaptable.

At that stage I'd only flown Tiger Moths, Ansons and Beauforts. I quickly converted on to Blenheims (short and long nose) and Beaufighters and then on to American twins such as the Maryland and Baltimore, the Bisley (a relation of the Blenheim) and then singles.

Conversion to singles was rather interesting. We were at an aerodrome north of Cairo at Landing Ground 267 on the Cairo/Alexandria road. They had one Harvard—but only one—so, to convert to singles, you'd go round with the instructor in the Harvard until you were landing it properly. Then they wouldn't send you solo in the Harvard because they only had the one.

They would send you solo in a Hurricane. Difficult to imagine anything less similar to a Harvard than a Hurricane. So you had to sit in the cockpit and get used to the feel of the controls. Then they'd send you off and say a few prayers for your safe return. Well, I got away with it.

I flew everything from a Spit I to a Spit VIII. I used to fly Tomahawks and Kittyhawks. They were the main singles that I flew.

Flying Spitfires was interesting. The Spitfire was extraordinarily sensitive in the fore and aft controls and for a ham-fisted taxi driver like me, who had been used to flying big heavy aircraft, it was a new experience altogether.

We had been warned about this and told to just hold the stick between finger and thumb and just move it gently—'Don't push it down or it'll do a slow roll and take off.' Anyway, the first Spit I flew was a Mark V. I managed to get it up and down again and took it round a few times. Later I actually ferried some Mark Vs. They were the standard Spitfire.

At one time there were eight Spitfires to be moved from an aerodrome called Bilbers on the eastern side of the Nile delta, up to Dekheila, which was an aerodrome at Alexandria. So eight were allocated to go out and fly them there. When we got to Bilbers I found, to my surprise, that the one I had been allocated was a Mark I—the very first Mark which had been made operational.

There was one considerable difference between the Spit I and the other Marks of Spit. In the Spit I, to get the wheels up you had to select 'Wheels up', as in the others, but then you had to pump manually! You had to change your hands over so that you had the stick in your left hand and pump the undercarriage up with your right hand.

The boys behind me said it was the most amazing thing they'd seen. There were a lot of sandhills in line with the runway and they said they saw this Spitfire going up and down like a yo-yo in line with the runway. That was me pumping the undercart and moving the elevator control at the same time. But I got the wheels up and I got the wheels down again and I delivered the aircraft successfully.

After that I converted to one four engined-aircraft—a Liberator. I delivered some of those.

But my great experience was delivering an American machine called a B 26 or Marauder. It was a very heavy twin-engined aircraft. It had a very light elevator control for a big aircraft but the thing about it was that it had a very high stalling speed. On the B 26A, the first version of the Marauder, you had to cross the fence, when coming in to land, at 135 mph and then hold it off. It was easy to land because it had a tricycle undercarriage and was fairly heavy. Once you got the main wheels on the ground it would stick on.

The B 26C had a slightly longer wingspan and it crossed the fence at 120-125 mph—but that was still a lot faster than anything we were used to.

I had one bad experience with a Marauder. I had to deliver it to Setif in the Algerian highlands. It was actually a Marauder OTU run by the Americans to train South African and RAF crews to fly these new machines. This was very late in 1943.

I took the Marauder off from somewhere on the Canal. It had belly tanks so I had quite a long range. I was able to fly to Tobruk (El Adem), where I refuelled, and then to Tripoli (Castel Benito). I was pretty low on fuel and it was getting on in the afternoon. I wanted to get to Setif because I knew a Hudson was coming through the next day and I wanted to get back. One of the disadvantages of the ferry job was that you could deliver a plane to some remote place and be stuck there for a week because there was nothing to fly back and you had to wait for somebody to pick you up. So I landed, taxied in and said, 'Just fill up the inners. I'm only going to Setif.' It wasn't very far from Tripoli.

I wanted to get away quickly so I dashed into the control tower and said I wanted to take the spare parts on to Setif that night. They knew me and gave me a quick clearance.

One of the things they always taught us was to be careful about the cockpit check. When you were facing the runway for take off, always do a check. On this occasion, because I was in a hurry, I failed to do that.

One of the techniques for landing the Marauder was to wind back the tail trim to make it squat down and hit with the main wheels first and make a safe landing. When I was coming in to land at Tripoli I wound back the tail trim to 13½ degrees tail heavy.

Harvard training aircraft.

To take off you have to have the tail trim 3½ degrees tail heavy. I did my cockpit drill but didn't stop. I glanced at the tail trim indicator but didn't read it properly.

About halfway down the runway the thing gathered itself like an athlete and just shot into the air like a rocket. I had to use all my strength against the control column to hold it in that position. But I managed to flick the throttle into 'Emergency' because I realised I was in desperate trouble.

They said, when I went through next time, that they'd never seen a plane take off like it. It was like a rocket with black smoke pouring out of the engines and going up at an angle of about 60 degrees.

I got it up to about 1000 feet, managed to get one arm off the control column, and desperately wound the tail trim forward into a controllable position. I reckon I came as close to being killed that day as I ever did.

Planes were often picked up from the squadrons. Sometimes we'd have to say, 'No! I don't think that aircraft is capable of being flown.' and we'd leave it. No trouble about this, they knew we'd take it back if we could. I never had to do it actually, but one or two of our blokes did. In most cases the only things wrong were things like holed tanks.

We didn't carry a crew so navigation was a matter of experience. Mostly we flew along the coast or there were dromes which were near the coast. We used to fly from the Canal area to Mersa Matruh and then on to Tobruk and Benghazi. Most of that was pretty well along the coast. We also used to land at the bottom of the gulf and go on up to Tripoli. When you had experience you could fly across the sea. I once delivered a Walrus to Malta.

I'd never flown a Walrus before. It was more like a World War I aircraft. It was an amphibian and had a pusher engine—the airscrew was behind the wing, not in front. It was also behind the pilot. To start it you had to crank it with a handle. When it got up to a high-pitched whine you pulled a toggle and it should start. It was an inertia starter. After three attempts to start it you had used up a bit of energy but you'd get it started eventually.

I picked it up from the navy—it was a Fleet Air Arm aircraft going to Malta for Air Sea Rescue purposes. I said to the young Sub Lieutenant (I was a Flying Officer at the time), 'If I've got to come down in the drink with this thing, what's the drill?' 'Well,' he said, 'the best thing to do is to stand back about a mile at a couple of hundred feet and steam in at about fifty-five knots'. 'Fifty-five knots!' I said, 'The thing will fall out of the air.' He said, 'No no. You can only do about seventy-five knots flat out.' 'Anyway!' he said, 'Don't try to do anything fancy! Just fly it straight into the water in a nose-up attitude. The floats and the hull will do the rest. It'll make a hell of a bloody splash. However, the machine will cope.'

The first day, after having all this instruction, the thing was so slow I only got to Mersa Matruh by nightfall. It was about 200 miles, at the most, from Alexandria.

I landed on the aerodrome there. The wheels of the thing came down below the hull, but not far below, only about a foot. I stayed the night there. Next morning they filled the tank and I took off. As I was doing the take-off run on the rough desert strip there was a hell of a bang and I realised one of the tyres had burst. I was just about unstuck so I dragged it off. Then I thought about what I would do now. If I tried to land with a burst tyre I would undoubtedly puncture the hull or break off the undercarriage. So I decided the only thing to do was land it in the drink.

There was a lagoon, a sort of enclosed area, near the aerodrome so I did what the young snotty had told me. I got the nose up and came in with a fair bit of power at about fifty-five knots at 200 feet and flew it down on to the water.

When he said there would be a hell of a splash it was the understatement of the century. A great fountain of water shot up. When it subsided, there was the Walrus, afloat, with water pouring off the wings.

I taxied it in and a bloke swam out with a rope. They hauled it in and I explained what had happened. They didn't have a tyre and neither did I. So they sent back to the Canal for one which came up on a transport aircraft. They fitted it and off I went.

I also took a Baltimore to Malta. The Baltimore was a beautiful aircraft to fly.

On one occasion I was detailed to go up from Egypt to Gibraltar as a passenger. There I was to pick up a Hurricane and lead a flight of Hurricanes from Gibraltar to a place called Shaiban in Persia (now Iran). It was close to the Russian border and the Hurricanes were to be handed over to the Russians for use on the Russian front [September 1943].

When I got to Gibraltar the Hurricanes were there but the pilots were not. The planes had come by ship and had been taken to the strip at Gibraltar. The pilots, all sergeant pilots, were coming from an OTU in England but had not yet arrived so I had to wait at Gibraltar for a day or so.

On the second evening I was invited by a friend to go with him to a naval mess for dinner. I went over to the wardroom and had a very pleasant dinner. Then the CO of a Fleet Air Arm squadron said, 'Can you fly a Beaufighter?' 'Yes.' 'Oh good! We've got a Beaufighter here that needs testing but I haven't got anyone to fly it.' These people were flying Fairey Barracudas. They had fixed up this Beaufighter and wanted it air tested so it could be flown away.

I reported to the Flight Office and, to my amazement, I found it was a Beaufighter Mark II. It had Merlin engines, which were in-line liquid cooled engines, whereas all the Beaufighters I had flown had radial engines.

Anyhow, I felt I couldn't back out now, I had flown with Merlin engines in Spitfires.

I had been taught, when flying Beaufighters Mark V or VI, to always lead with the starboard throttle when taking off as the aircraft tended to swing to the right. But the Mark II's swung the other way, as I now found out.

I started off, leading with the starboard throttle, and did a very smart ground loop in the middle of the aerodrome, just missing two Wellingtons and a Liberator which were parked there. I had realised what I had done, but too late to correct it. A fellow came shooting out in a jeep. He called out 'Everything all right, old boy?' I said 'OK old boy.' Then I taxied back, took off, did the air test and landed.

• Flying Conditions in the Western Desert

The airfields or, to be more precise, stops—because they were only strips with few facilities—provided many problems for the pilots. The tricky flying conditions also made navigation difficult.

Jack Stronach experienced a novel situation on one strip on which he made an emergency landing:
Whilst serving with 117 Squadron RAF (Hudsons), on 11 February 1943, I took off from Marble Arch in Libya, which was our base at that time, to fly to Castel Benito, the aerodrome for Tripoli.

After flying for two and a half hours we flew into a sandstorm which grew in intensity. We were now over a range of low mountains and could just see the ground through the storm. We saw a road and decided to follow it. After a short time, and much to our surprise, we found we were flying alongside an aerodrome. I turned and made my approach through the storm and realised we were landing into a gale. Just after the wheels touched the ground the aircraft came to a stop.

Immediately after I landed, a Spitfire was flicked over on to its back while trying to take off and a Dakota had to abort its take-off when it was nearly turned on its back. I realised how lucky I had been to make a safe landing.

The next day we were unable to obtain a weather report for the Marble Arch area. In the afternoon I decided to give it a go and return to base. The clouds were down over the mountains so I decided to fly north to the coast and then east, along the coast, to Marble Arch.

We had been flying for about an hour when we ran into a bad sandstorm. I came down to 1000 feet so that we could keep the black strip of the coast road in sight. I told the navigator to keep the road in sight and to guide me port or starboard.

We had been flying for over two and a half hours when we realised we had flown over our base at Marble Arch without seeing it. Things were now starting to get serious. The sun was starting to set and I had to get on the ground before darkness fell.

Luck was with me again. The sandstorm started to subside and we could see the ground. We realised we were in the area of Agedabia and that there was an aerodrome in the area.

We found the aerodrome—the boundary was marked with 44-gallon drums. The light was fading fast as I was making my final approach.

I thought, 'That's strange!' There was a row of 44 gallon drums across the middle of the aerodrome. I skimmed the wheels over the drums and put the Hudson on the ground. I soon ran past the boundary, missing everything, and then taxied back to where there was an American Dakota and its crew.

We went over and asked them why the drums were across the middle of the aerodrome. They said I had landed on the right side of the drums because the other side was heavily mined. Luck was with me again.

That night we tried to sleep in the aircraft. We were only dressed in shorts, shirt and socks. When the sun goes down in the desert the heat leaves the sand very quickly and the nights become very cold. It was like trying to sleep in a refrigerator. Also, we had no meal that night.

When dawn broke I flew back to our home base and we had a breakfast of bully beef and dogs' biscuits—which we enjoyed.

Jack Stronach [above] and his photos: his desert accomodation [below] and in front of his Hudson.

Ron McCathie (a navigator) tells of his problems:

Ours was one of the few RAAF squadrons in the Middle East. Both air crew and ground crew were four-fifths Australian. The remainder were lovely Pommies. The squadron was equipped with Baltimores at the time.

The ground crews were superlative. The squadron won the serviceability record in the Middle East two years in a row. This was based on the number of serviceable aircraft available for operations week by week during the year. How they managed this I will never know.

In the summer the temperature was 110 degrees in the shade with sandstorms blowing at forty miles an hour and no aircraft under cover. In these conditions no-one could touch the aircraft on the ground without wearing gloves. Contrary to regulations, we flew without shirts.

Reverting to these periodic sandstorms; they could be so bad that visibility was virtually zero. So much so that it was necessary to rig lifelines from your tent to the mess. It was possible to endanger your life if you did not use them.

Of course, cooks were having a bad time too. Having crawled out of your bunk, where you had spent the day with a sheet over your head, all the poor cooks could produce was William Angliss bully-beef fritters with sand gravy. It was hard on the teeth.

What puzzled me at that time was the phenomenon of having a wind of high velocity and no sand storm. The explanation came later. It all had to do with the barometric pressure. If the wind was at forty miles per hour with a high pressure there was no sandstorm. On the other hand, if the pressure was low, the sand on the surface was destabilised and, in part, defied gravity, got itself up in the air and made life miserable. I think the Met blokes referred to this phenomenon as surface tension.

If one of these storms arrived during your absence on a sortie, it was quite possible to get back to the strip and see the wind sock but nothing else. Twenty feet up the visibility was perfect, but, under that, nothing.

The poor old pilot would come in knowing he had twenty feet to go but could see nothing at all. He was left to feel for the ground with the seat of his pants, hoping no clot had parked any motor transport where he thought the strip should be. There could be no thought of aborting the landing. The nearest alternative landing strip could be 300 miles away and the fuel supply insufficient.

ABYSSINIA
JULY 1940 TO MAY 1941 AND LATER

In 1936 the Italians conquered and then occupied Abyssinia (now Ethiopia). Italy still occupied Abyssinia when it entered the war at midnight on 10 June 1940.

The Italians' presence in Abyssinia and, indeed, their advance in July 1940 to Kassala and Gallabat in the Sudan, was tolerated until December 1940. Steps were then taken to evict the Italians from the Sudan.

To do this, Wavell employed the 5th British-Indian Division. These regular troops could not be spared for penetration into Abyssinia itself. It was decided to mainly use the Patriot movement to make the Italians' position in Abyssinia untenable.

Brigadier Sandford and one Sudanese battalion, together with Colonel Wingate and a small number of British officers and NCOs provided the nucleus of this force. Ultimately, Orde Wingate led the Gideon Force consisting mainly of irregular patriot forces. He was accompanied by the deposed Emperor, Haile Selassie.

In January 1941 this force entered the enemy territory. It was matched against several thousand Italian and Abyssinian troops.

Supplies were paid for in 'Maria Theresa' dollars and the local tribesmen were paid in the same currency, for giving their support.

On 5 May 1941 Haile Selassie and Wingate entered Addis Abbaba—the Italians having surrendered the city and the country.

Ron McCathie tells of his participation in the aftermath:

The Italians had surrendered in Ethiopia. But the British government was still subsidising the insurgent tribesmen who had played a part in bringing this about.

It seems we were obliged to deliver a quantity of Maria Theresa silver dollars each month. These dollars were the only currency readily acceptable in those countries bordering the Red Sea. They had been minted in England for about one hundred years. They were about the size of our old five-shilling piece. They were heavy.

For our purpose, the coins were packed in canvas bags weighing twenty pounds each. Army Intelligence would nominate time and place, we would be briefed (the maps of the area were good), load up twenty to thirty bags in the bomb bay and go off on our four hour flying mission. The track took us up the full length of Lake Rudolf and into Ethiopia.

Two or three hundred tribesmen would be scattered about the designated drop zone waiting for their handout. Three or four low level runs would complete the mission.

The canvas bags exploded on impact. The ensuing scramble for the scattered coins was hilarious. There did not seem to be any ground rules, it was simply every man for himself.

Occasionally some greedy fellow could be seen, arms wide apart, trying to catch a whole bag for himself. When sometimes successful, the result was shattering. Apparently they were never taught anything about dynamics. Twenty pounds with a velocity of 120 miles per hour has a punishing impact.

The poor skipper and the hard-working gunner pushing the bags out could see nothing of the fun on the ground and depended on the navigator up front to provide a running commentary through the intercom.

Imagine ten or more VFL football rough-house rucks all going on at the same time. Legs and arms in every direction. Then add the flowing robes (gallabeas). It was better than a circus and left the crew in fits of laughter most of the way back to base at Nakaru.

An interesting sidelight to operations over Ethiopia was the sidearms issue as related by Ron McCathie:
Among the gear issued to each of us on joining was a .45 Colt revolver. No doubt they had been in store since the end of the First World War.

Out of the blue the Jerries issued an edict—'Any soldier or airman captured wearing a .45 would be shot.' No mention of a court martial.

Within a few days of the announcement the Colt .45's were withdrawn and, in their stead, we received .38 Smith & Wessons. It was all very confusing to those of us entitled to wear sidearms.

Subsequently we learned that the Colt .45 was designed for trench warfare in the 1914-18 war. It had a soft nosed bullet that blew a man apart at ten feet—before he could reach you with a bayonet. On the other hand, the .38 Smith & Wesson fired a steel-jacketed bullet that would go straight through a man without blowing him apart. The Germans apparently thought this a more civilised way to shoot each other at close quarters.

OPERATIONS FROM MALTA SEPTEMBER 1940 ONWARDS

Malta was strategically located between Sicily and North Africa. It controlled access to the eastern and western parts of the Mediterranean. It had a fine harbour at Valetta. Thus it was hotly contested. The British fought desperately to hold it while the Italians and Germans fought equally hard to subdue and conquer it.

On 19 September 1940, Flight Lieutenant Tich Whiteley flew into Malta with four Glen Martin Maryland bombers. These aircraft were responsible for the reconnaissance of Taranto which led to an eventual attack by Fleet Air Arm aircraft in a very successful sortie.

Thereafter, other aircraft used Malta as a stopping off place to attack convoys.

Jim McSharry tells of one of these sorties:
This happened in August 1942. I had been wounded in operations in June 1942 and, at this stage, I was still recuperating.

While I was out of action my navigator had also become ill with the local brand of dysentery. This left my two wireless air gunners without any crew. They were two New Zealand boys named Brown and Wilkinson. Brown was a railway fettler and Wilkinson was a PMG linesman.

A maximum effort was called for on this day to attack a big convoy near the Greek coast. It was on its way to reinforce Rommel in North Africa.

The CO of 217 Squadron (Gibbes) ordered Brown and Wilkinson to do the trip with a South African pilot named Ted Streber and his English navigator (Dunsmore) who had lost their two gunners. So my two boys made up a scratch crew.

Their aircraft was shot down in the attack on the convoy. The aircraft was a Beaufort I which would not fly on only one engine. One engine was put out of action so Streber had no choice but to ditch. He was a good pilot and did a copybook ditching and they all got out, unharmed, and into the dinghy.

Shortly after, they were menaced by a fighter aircraft which was apparently lining up to strafe the dinghy. One of our Beaufighters, which had been supporting the strike aircraft, came along and shot down the enemy fighter into the sea.

About one hour later, a Cant three-engined seaplane came along, landed near the dinghy, picked up the four crew members and made them prisoners of war.

They were flown back to the Italian squadron's base near Corfu, arriving there late in the afternoon. They were treated well by the Italians. All four were accommodated in the officers' quarters even though my two gunners were NCOs. They all had dinner in the Officers' Mess. Everything was jolly and the Italian wine flowed freely but the boys were very careful in case the Italians were trying to get them drunk and get information from them.

Nothing untoward happened and they went to bed in their rooms in the officers' quarters where they were guarded all night.

Next morning they were given breakfast and then loaded on board a Cant seaplane which was being flown back to Taranto, in Italy, by a crew due for home leave.

The Italian crew consisted of a pilot and a navigator, who both sat at the front of the machine, a wireless operator, who sat behind the pilot, and a guard, who was a ground staff corporal, with the job of guarding the four prisoners. They were told they would be taken to Taranto for transfer to a POW camp.

They were heading for their turning point, due west from Corfu, where they would turn north west for Taranto. Conditions were quite rough. The plane was flying quite low and the guard became air sick.

At a strategic moment, Sergeant Wilkinson, the PMG linesman, got into conversation with the Italian wireless operator, who could speak a little English. Wilkinson got the chap to lift his chin up by pointing up high as he asked a question. When he did that, Wilkinson gave him a beautiful right uppercut and knocked him out.

He picked up the man's body and threw it into the lap of the guard, who was being airsick. He seized the guard's pistol and held up the rest of the Italian crew. The four men on our side quickly took the revolvers of the other crew members—they had been stowed in a locker. So now all four were armed. All of the Italians, except the pilot, were then ordered into the back of the machine.

Lieutenant Streber got up in front with the Italian pilot and they flew the Cant across the mouth of the Gulf of Taranto, keeping low and following the coastline at a respectable distance. They proceeded until the navigator recognised the cape at the southern end of Italy. Then they flew across the Strait of Messina until they saw the coast of Sicily.

They kept low and well out to sea in the hope they wouldn't be seen by enemy aircraft. When they got to the southernmost point of Sicily they knew they would only have to fly west of south for a short distance to find Malta.

As they approached Malta, the inevitable happened. They were attacked by a flight of Spitfires. However, they had anticipated this, and the two New Zealanders took off their shirts and singlets and waved the white singlets from the twin gun positions in the waist. The Spitfire leader saw the singlets and got the message.

In the meantime, Streber was waggling the wings, which is a generally recognised signal for surrender. The Spits circled round but they didn't fire any more and Streber got the Italian to bring it down in quiet water near St Paul's Bay.

For some time the Italian pilot had been pointing, in an agitated manner, to the fuel gauge and, as they were taxiing in to the shore, the Cant ran out of fuel and had to be towed in.

They were brought ashore and a big crowd of Maltese had gathered. They started displaying their enmity towards the Italians by spitting at them, cursing and threatening them.

Lieutenant Streber took out the revolver he had taken from the Italians and kept the Maltese at bay, telling them the Italians had treated them decently while they were prisoners and they were going to treat them decently.

The whole thing caused quite a furore. The RAF had acquired a Cant seaplane—with a few Spitfire bullet holes in it, four Italian prisoners of war and the return of their own aircrew.

The outcome was that Lieutenant Streber and PO Dunsmore were awarded DFCs and Sergeants Brown and Wilkinson were awarded DFMs.

THE WAR IN THE SUDAN

During late 1940 and throughout 1941 the Sudan had been a base for British forces and irregulars invading Abyssinia. Britain had maintained a dominant interest in the Sudan since 1898. From the early 1920s, however, there had been a rising tide of nationalism but Britain had managed to maintain her position and avoid widespread rebellion by isolating the south. This had been done by control without occupation, that is, by using aircraft to quell tribal disturbances. This had been a continuing process.

Thus, it had been necessary to station aircraft in Khartoum, the capital, and the outbreak of war required the continuation of this policy. In 1939, and for a considerable time thereafter, this force was comprised, in the main, of Blenheim and Wellington aircraft.

Jack Stronach tells of some problems he encountered while flying in the area:

I had to go to a place called Wadi Halfa in the Sudan. I was detailed because I was a fully trained Hudson pilot and most of the other pilots were still on the training program. I was to search for a Blenheim that had come down in the desert some distance from Wadi Halfa.

I took off early in the morning and flew down the Nile River at about 1000 feet. The flight lasted four and a quarter hours

As I approached the aerodrome to make my approach I came into a heat mirage with all the ground shimmering as it does when you drive along a bitumen road on a hot summer's day. It was difficult to figure out what was what. The heat was terrific. I was told later that it was regarded as the hottest place in the world.

Then, on making my approach, I could see what looked like a big row of posts across the middle of the aerodrome. On my final approach I could see the posts were twenty-five to fifty feet high and in a straight line across the aerodrome. As I came lower I started to realise that they had been put there to give the pilot some idea of where the ground was in the mirage.

I landed successfully, although somewhat steeply, despite the poles.

The Blenheim was later found but the crew had not survived.

A Polish fighter pilot came in and landed while I was at Wadi Halfa. He was smuggling gold which he had in a bag. He buried the bag under one of his wheels so that he could find it when he went on to Cairo.

When he returned the next morning to continue his flight, he found they had moved his aircraft. I guess the gold is still there, hidden by the featureless desert sand.

OPERATIONS IN IRAN AND IRAQ
SEPTEMBER 1942 – DECEMBER 1942

The Allies needed the support of Iran and Iraq for three main reasons: (1) There were rich oilfields in the areas under their control; (2) The LendLease supply line to Russia lay through the area; and (3) The Middle East-Persian Gulf-India supply line had to be preserved. But support relied on both protection and pressure to maintain Allied interests.

Accordingly, one brigade of the British 50th Division was sent to Iraq. This move was to counter the push by Rashid-Ali, the Shah, who wanted to ally Iraq with the Germans who were pushing down into the Caucasus.

When 454 Squadron RAAF was reformed in September 1942, a ground staff party of 500 personnel was sent to a location near Mosul in Iraq. The squadron was equipped with Blenheim V aircraft.

Squadron Leader Ron McCathie, a navigator with the squadron, gives some insight into various aspects of life there as he tells of his meeting and experience with the Russians. He also gives a humorous sidelight on the task of map-making:

• Coexistence with the Russians

The unit was ordered to move on to Kalamorgie Aerodrome, two miles out of Tehran. Much to our surprise, it was in the hands of the Russians—tight security and guards everywhere. I never did find out how this came about.

There were delays and difficulties before we were allowed to move in with our vehicles. Our aircraft had already arrived. The Russians were an unfriendly and surly bunch of bastards who made living together on the same airfield as difficult as they could. We were with them for a total of five months and there was never any fraternisation.

The Russians used Kalamorgie as a staging post for the aircraft being supplied to them by the Americans. These were flown across the Atlantic to the Azores for a fuel stop. They then went on to Lagos or Takoradi on the West coast of Africa and then on to Khartoum, Cairo and Basra, where they were handed over to their new owners. Our allies would not sign for them and take delivery if there was one item missing from the checklist—even down to a shifting spanner from the tool kit.

The aircraft were mainly Bostons, Baltimores and Hudsons. They would fly up to Tehran from Basra for a night stop and fly across the mountains next morning to Azerbaijan where the Germans were close to taking possession of the Baku oilfields. Some of them would be operational within twenty-four hours. The delivery rate was about five aircraft each day.

One morning a Boston took off and the pilot retracted his wheels just a fraction too soon and made a belly landing. The aircraft was a mess but he escaped with only a broken leg. The Russians put him in hospital and put his leg in splints. The next morning we heard shooting on the far side of the field. We eventually found out that the poor fellow had been executed for his misdemeanour.

They were tough customers and, no matter how hard we tried, we never managed to make friends with them.

• Map-Making

Aerial reconnaissance in Southern Persia was not a dangerous occupation except in the event of a forced landing. Each of our Blenheim aircraft, at 454 Squadron, carried a week's supply of hard rations and water, together with three .303 Lee Enfield rifles, 1000 rounds of ammunition and several Mills bombs.

Our instructions were that, in the event of a forced landing, on no account were we to leave the aircraft. We were to prevent capture at all costs until a rescue party arrived.

We were told that, if taken by the tribesmen we would be handed over to the women and would finish up with squeaky voices for the rest of our lives—not a happy prosect. This was supposed to be a practice that had been going on for thousands of years. The theory behind it was that, if you limited your enemy's breeding capacity, there must surely be fewer of them to contend with as time went on.

GENERAL MARITIME RECONNAISSANCE JANUARY 1943 TO ARMISTICE

Anti-submarine patrols, long-range penetrations to the harbours of the Aegean Islands, diversionary operations, convoy cover and convoy attack were some of the activities engaged in by 454 RAAF Squadron and by RAAF members of RAF squadrons during this period.

The following episodes give some picture of the experiences of aircrews on these tasks.

Squadron Leader Ron McCathie tells his story of some of the squadron's activities:

In January/February 1943, 454 Squadron was equipped with twelve Baltimore light bombers. One flight of six was fitted with extra fuel tanks for long range reconnaissance, giving an endurance of about seven hours. The other flight of six was fitted out as bombers.

Crews could be detailed for either flight, depending on the job in hand. I preferred the long range general reconnaissance work—it was all dead reckoning navigation and a navigator's dream.

One job consisted of providing close support for convoys between Malta and Alexandria by sweeping in front of them looking for submarines. The other task was preventing the Germans from supplying their garrison in Crete.

• Convoy Escort

McCathie continues:

On one occasion we had to intercept an inward-bound Allied convoy at extreme range as the Malta people handed it over to us. The intercept involved a flight of 600 miles. Visibility was down to about three miles and I was biting my fingernails as the ETA for the intercept drew close. All of a sudden, the ships loomed up with scarcely sufficient time for us to fire off the colours of the day. The navy were not very good at aircraft recognition.

There were eighteen ships in the convoy, arranged in three lines of six ships in line astern—about two miles between the lines.

When flying at 500 feet over the ocean, navigation by drift measurement can be incredibly accurate. The convoy was only four miles wide and we had 'hit' the starboard leader.

At that range we could only escort them for one hour before handing over to our relief and heading back home.

• General Reconnaissance

Ron McCathie describes what was involved:

There are dozens of Greek islands in the Aegean Sea. Every one that had port facilities was flown over each day. We flew at an altitude of two to three hundred feet from island to island. When about five miles from the next port we'd climb to 10 000 feet to avoid the light flak. If there were no ships in the port, we'd take an oblique photograph to confirm the fact at our debriefing and dive down again to the water and our next island.

If enemy shipping was sighted, we flew over the ship to photograph it with the vertical camera to allow identification and radio silence was broken. Beaufighters would be sent in from Cairo.

On these occasions the people on the ground reacted most disagreeably. The crews were always happy when an inspection proved negative.

German fighters based on the Greek mainland and Crete patrolled the Aegean to prevent this inspection. But, in six months, we only saw one. We dived, and at one hundred feet, 'went to sea.' Jerry followed for about five minutes and then gave it away.

With us so close to the water it was dangerous for him to dive on our weaving aircraft with four .5 inch guns in the top turret; apart from which, single-engined pilots prefer not to fly over water. Then again, he might just have been short of fuel. Five minutes after Jerry quit we turned round and went on with the job in hand.

From Benghazi to Cyprus about seven ports would be covered. After stopping overnight at Nicosia we'd return to base next morning by a different route. The trip took about five hours.

It was difficult to realise a war was going on in such beautiful surroundings.

• Photographic Reconnaissance

Ron Gardner:

I was on a photographic Reconnaissance Unit flying Spitfires. I was based at Matariya, near Heliopolis, in Egypt. We used to fly from there to our forward detachments. One of these was at Tocra, forty miles east of Benghazi and another was in the northern end of Palestine. With our long-range Spitfires we could cover the whole of the Eastern Mediterranean from these detachments.

Tocra was a tiny little airstrip on the coast with a detachment of four Spitfires. It had a macadam runway and we were just a few miles from Benghazi. From there we used to cover the area from Crete to Greece and right through to Rhodes and the Dodecanese Islands.

Spitfire.

My longest flight in a Spitfire, at this stage, was five hours and ten minutes. We could cover many miles and at height we felt we could see more than any other person could ever see.

PRU Spitfires were specially modified. When a new mark of Spitfire came off the production line they ironed out all the teething problems, which always came with a new mark of aeroplane. Then they grabbed about twelve off the production line for PRU modification.

They took out all the armour and the wing was modified to make the whole space, from the main spar to the leading edge, one gi-normous fuel tank. We could take those Spitfires out for well over five hours.

To get the maximum range we had to go fairly high, so all the armament was taken out and we carried the bare minimum of radio equipment. Right behind the pilot, where the radio equipment had been, they installed two cameras, each at a slightly different angle to the vertical, so that we could take a line of overlapping pictures. Cameras ranged from twenty-four to thirty-six inch focal length. The latter could just fit into a Spitfire.

We were particularly looking for shipping or troop movements or the making of aerodromes. We could fly right on the deck with an oblique camera in the fuselage looking for any movement at all.

By making daily observations it was possible to detect small changes which might indicate some enemy activity. The photo interpretation people could tell what was being made in a factory from photos of the scrap and refuse around it.

One of the most satisfying times was when I spotted and photographed a German convoy of four destroyers, a cargo ship and some E-boats. That was a great thrill.

The beauty of reconnaissance was the variation in the type of targets. The Corinth Canal was one such variation. It saved hundreds of miles for ships and its narrowness and steep sides protected shipping in it from attack so it was used as much as possible by the enemy. We used to monitor the usage.

In 1943, the PRU Spits provided 90 per cent of the information on shipping and troop movements and port congestion for our Intelligence.

When the army started to move into Corsica, Sardinia and southern Italy, one of our chief objectives was to trace the movement of the German Storks, a beautiful little aeroplane. They did 100 mph with a pilot and one passenger. They would transport food, soldiers, anything at all. They flew at night from Greece to the islands—particularly Crete.

We had to look for any activity by the Storks because they were beginning to hurt. Up to a hundred were operating each night and this was causing a lot of problems. The important thing was to find their landing places. When they made their run they would take a wounded soldier or anything back to the Greek mainland. In this way the Germans kept open a small but important supply line.

I had a particular success in this area. I could see a lot of activity on the ground and it turned out to be a new Stork landing ground which disappeared under a copse of trees. I managed to get some evidence of it on photographs which, when developed, showed a lot of German activity around Navarino Bay, probably associated with a main Stork equipment area. The Beaufighters went in and blasted the hell out of it. My work did some good.

It was an amazing area in which to work. At one time I was over Crete at 28 000 feet. Looking backwards I could see the forbidding, dirty, dusty coast of Africa behind me. Straight ahead, to the North, I could see the snow capped mountains of Greece. To get the two contrasts and be over enemy territory at the same time was quite something.

Another feature of the type of flying I was doing was that we were lone wolves. We never flew in company. We always took off on our own, did the trip—hundreds and hundreds of miles doing our own navigation—and never saw another soul till we landed at base.

I had one incident which sticks in my memory. I was over Rhodes and came down out of the cloud to be immediately boxed in by the German ack-ack. I tipped the Spitfire over and went in and out like a bat out of hell. I later found that a Mosquito, who shouldn't have been in the area, had been over half an hour before me and woken up the defences.

All in all, I did forty-five trips in the Middle East.

Ron Gardner had three prangs. The second was in a PRU Spitfire during operations over the Egyptian desert: This accident was quite severe. It was in a really old cranky Mark V Spitfire with the tropical filter under the nose.

I had taken off at about 11.30 a.m. and had been flying for about three hours and was at 28 000 feet when the motor started to play up and then seized. It must have developed a glycol leak after I took off and I didn't become aware of it until the engine started to play up.

I glided to within twenty-two miles of Ishmailiya and put it down in a good belly landing. There were bits of Spitfire over about half a mile of desert. I wasn't hurt except for a slightly jarred back.

I walked out of the desert dressed in full flying gear— two pairs of socks, flying boots, battle jacket and the rest—just carrying my maps and my Verey pistol. I walked the twenty two miles across the desert to the RAF drome at Ishmailiya—it took seven and a half hours. I was pretty lucky.

• **Corsica**

Corsica was occupied by the Italians when they entered the war and it remained under Italian/German occupation until October 1943, when it fell to the Free French under General Giraud and was then available as a base for Allied maritime reconnaissance.

Flight Lieutenant Jack Robertson was a pilot with 14 Squadron RAF: Our crew consisted of first pilot, second pilot, navigator, wireless operator, mid-upper gunner and tail gunner.

14 Squadron was used for low level anti-shipping reconnaissance in the Mediterranean. It had been originally operated Blenheims but converted to Marauders in late 1942. Before I got to the squadron it had been doing a little bombing—mainly skip bombing—and dropping torpedoes.

Thank goodness I joined after that. Dropping torpedoes was a somewhat dicey experience. You had to fly directly towards the target, straight and level, for two or three hundred yards, at a height of fifty feet just before releasing the torpedo. That is not recommended for your health.

The Marauder was very good in that it could be skidded from side to side so that you had some evasive capacity. I was told that the training was often more scary than the actual operation.

When the squadron first started torpedo dropping, they used to go out in the desert, have the outline of a ship marked out with oil drums and make mock attacks. It was not until an aircraft lost its rudder during one of these violent skidding attacks and flipped over that it was found that the tail fin needed strengthening.

However, the squadron did have a very good record of ship sinkings using torpedoes.

When I joined, the squadron was doing the normal anti-shipping coastal reconnaissance. We would go out, get down to fifty feet over the sea, and fly to some predetermined point off the coast. We'd run along the coast for a few minutes, then go out to sea and come in at another point and repeat this for the duration of the patrol.

On one occasion, when we were stationed on Corsica, I was on the way out to start the patrol. I had to skirt a number of sea fogs which extended from sea level to about 200 feet. These made it difficult to get exactly on our start point, but just as we got there and did a tight turn to run along the coast I found us heading straight towards three Arado float planes at the same height and coming at us head-on.

Normally, the Marauder could outstrip any opposition at sea level because it could get up to 300 mph plus for about five minutes before the cylinder heads melted. But we were caught a bit flat-footed at about 200 mph with the Arados too close. They flashed past, turned beautifully and came in on our side.

Marauder flying off the Corsican coast.

Two photos from Jack Robertson's reconnaissance flight immediately following the attack on the Italian liner *Rex*. *Left:* Approaching the stern of the vessel which can be seen still smoking in the lower left of the photo. *Right:* Much closer, the ship is listing severely, the deck facing the camera.

Normally, the w/op would be in the astro dome to give fighter control, the theory being that he would give instructions over the intercom while the navigator would reach across and use the w/op's key to send out FGQ, the aircraft identification and position. FGQ meant 'I am being attacked by a fighter.'

On this occasion, unknown to anybody, when the navigator reached across to send the message, he accidentally pulled the w/op's intercom out of the socket. So we had no fighter control. All I could hear was the two gunners belting away. I immediately headed for a cloud bank some 10 miles away, corkscrewing between 20 and 200 feet.

I heard the rear gunner stop firing and I thought, 'He's had it.' Then I heard the mid-upper stop and thought the same. I thought, 'I'm in trouble!' Then I heard the Lancashire voice of the mid-upper gunner saying, 'It's all right Robbie! Stop panicking! They've been gone for five minutes.' I was most relieved.

A lot of people malign the Marauder. It was known as the Flying Prostitute—especially the earlier models we were flying, which were the short wing variety. It got its nickname because it had no visible means of support. It had a very poor single engine performance.

One of our pilots lost an engine at Venice while at fifty feet and it took the whole 400 miles back to base to get up to 1200 feet, the circuit height we had adopted.

The normal air force circuit height was 1000 feet but, because of the very high landing speed of the Marauder, we used to make our approach at 150 mph on the final cross-wind leg. We did a fighter approach, dropped our wheels on the downwind leg at about 185 mph, reduced to 150, dropped flaps and touched down at somewhere between 110 and 120 mph

I didn't sink a sub but I was responsible for finding the largest Italian passenger vessel sunk during the war. That was the Italian liner *Rex*, of about 44 000 tons, which had been holed up in Trieste.

Rex was a beautiful passenger liner. It had won the Blue Riband before the war and, of course, the Italians were very proud of it. According to our information, the Germans wanted to move it out of Trieste and hide it somewhere in order to use it later as a blockship across the entrance to Trieste harbour. I don't know whether this was true or not, but high-level reconnaissance of Trieste showed that the *Rex* had been moved out of the harbour.

Our squadron had the job of watching the coastline of the Adriatic. We used to survey the coast from a distance of five to seven miles at a height of fifty feet. We suspected that the *Rex* was somewhere near the Istrian Peninsular, at the head of the Adriatic Sea.

After a few days of searching, interrupted by very heavy thunderstorms which sweep from the Austrian Alps down the gutter which the Adriatic Sea really is, I went out and just happened to have the luck to find her anchored offshore, about forty miles from Trieste.

Jack Robertson's crew and Marauder bomber. *Top left to right:* F.O. Ingham (Air Gunner) F.O. Robertson (Pilot) F.O. Duffell (Second Dickie) *Bottom left to right:* F.O. Hanks (Navigator) F.O.Harris (Air Gunner) Flt. Sgt. Murray.

I sent out a sighting report and, when I turned to get closer to photograph her, I found I had turned into a little box of two destroyers and two armed merchantmen. I can tell you I wasn't so much interested in taking photographs as with getting myself out of trouble.

A strike by some rocket Beaufighters had been laid on and I was instructed to brief them at a forward base on the Eastern coast of Italy, near the front line. I had to land at a small aerodrome which was suitable for fighters and Beaus but small for a Marauder, which needed a pretty long runway. It was a pretty dicey landing.

I expected that, once I had briefed the crews I would be finished. However, they sent off the squadron of rocket Beaus with an escort of anti-flak Beaus and instructed me to follow a quarter of an hour later. I was to go without escort and photograph the results of the attack.

Fortunately, the rocket Beaus had done a good job and there was no sign of any enemy defences.

I learned my best lesson from a very experienced mid-upper gunner whom I had inherited from another crew when I went to command my own crew. He was an old man of twenty-eight, married with two children, but a damned good gunner with a lot of commonsense.

I sighted some German E-boats off Marseilles and came in to allow my second pilot to take photographs for fuller target identification for the rocket Beaus.

There was the usual exchange of angry shots and then a voice came over the intercom. It said 'Robbie! I've got a couple of things to say to you. Number one. When you approach the target you don't come in from dead astern because you give them a no-deflection shot. You approach from the beam and make it difficult for

them. Number two. You tell your crew what you're doing and don't put their lives at risk.'

Jack Robertson tells of a humorous episode when he was stationed there:

We were a very happy squadron, especially as we served in Corsica, Sardinia and in Italy.

I remember an evening when Florence Desmond came out with an ENSA concert party. She did her show and then insisted on seeing the area in which we were quartered. We were in an old vineyard a couple of miles away from the aerodrome. We had tents all round this big block of ground and, right in the middle, was our very naked toilet facility, such as it was. Florence was taken short and everyone was ordered into the woods while she attended to her needs.

• Crete

George Gray tells of a reconnaissance trip to seek out Axis supply vessels:

I did one long trip—from Mersa Matruh to Crete—in which we had complete cloud cover above us the whole time. We were at about 1500 feet and above was a steady stream of Ju-52s with fighter cover above, bringing petrol over from Crete for the Afrika Corps.

The long-range fighters went up every so often to get them but we were ordered to stay down low and look out for submarines and shipping. We had to report on the transports which were aiding and abetting the airlift above.

I remember that particular trip—my last in a Blenheim—because I hit the point of return bang on the nose after five or six hours airborne at low level, coming in at 15 degrees to the coast and only using dead reckoning of the wind. I was quite proud of myself.

• Convoy Attacks

From the outset of war, as Churchill pointed out, 'we had to drive German commerce off the seas and stop all imports into Germany' (The Second World War, Vol. II, p. 19). All ports were subject to blockade.

Ships headed to or from neutral ports were subject to this blockade. Spain was a neutral suspected (with good reason) of running the blockade from time to time. Its government was pro-Axis.

Jack Robertson tells of some of his experiences on general reconnaissance duties:

The Germans were importing iron ore and other goods from Spain. The blockade runners, ships of perhaps 3000 to 4000 tons, would sail at night.

Sometimes the Wellington squadrons would pick them up, but normally they'd hug the coastline and hop into one of the smaller ports along the route from Barcelona to Marseilles at daylight. Occasionally they would come out during the day and try to make a few extra miles of their trip and we'd try to pick them up.

If we did find them we'd get a message back to base and out would come a squadron of Beaus to do them over.

It was frustrating to see these ships and yet not be able to attack them. But the function of the reconnaissance squadrons was to find the ships. It was the task of the specialist ship-busting squadrons, with Beaufighters armed with rockets and cannon, to attack them.

George Gray describes an exciting full-scale attack on a large convoy:

It was a magnificently organised strike. 454 Squadron featured in two different ways.

As a general reconnaissance squadron, we were flying long-range Baltimore IIIs with petrol tanks instead of bombs, giving us six and a half hours of flying.

We covered the Aegean Sea, the north of Crete and Southern Greece and the hideaway spots for the reinforcement of Crete which, at the end of the war, held something like 150 000 troops. In addition we patrolled the islands of Rhodes, Leros and Cos as well as the other islands which were the ring of outposts set up to hold out Churchill's much vaunted strike into 'the underbelly of Europe.' He was spouting that sort of talk in late 1943/early 1944 while D-Day was approaching.

Whether it was a planned exercise or not, the Axis worked into our hands by sending a convoy from Greece to reinforce Crete. It consisted of three merchant vessels, four destroyers and seven or eight escort vessels (motor torpedo boats etc.).

A Wellington of 38 Squadron picked it up at night and from then on 454, in its general reconnaissance role, sent aircraft at about one and a half hour intervals—

eight in all—to help keep tabs on where it was going. It ended up in the Aegean Sea.

A South African crew from 454 did the first tracking shift, even while they were jumped by Me-109s and Arado seaplanes. They kept going in and reporting the convoy and coming out at zero feet. The convoy came through the straits between Crete and Greece—between Kithira and Andikithira.

Usually there were duty pilots in 109s waiting above. They were doing their circuits as we came in and out. We had to try to time the run in so that they were getting petrol at Maleme as we went through.

The next crew—Ennis was the pilot and Moreton the navigator—was attacked four or five times by a 109 which kept coming back. He evaded the fighters as he was doing about 300 mph and, unless (because he was on the deck) a 109 dived on him, he could hold him off. He received a DFC.

Other 454 crews went in at intervals. We only lost one crew on this operation.

In the meantime, because of our experience with Baltimores and bombing, we were sent, with four other 454 crews, down to a strip near Tobruk to provide the leaders for a strike force of 17 South African Baltimores (15 Squadron) who were fitted to carry bombs. We were to lead them because we trained most of the S.A. crews at OTU.

Reconnaissance photo of a schooner off the Riviera coast. Had this vessel not been powered by sail it would have been attacked as it was believed that the Germans were only supplying fuel for use on their behalf.

We went off in two boxes of nine to rendezvous with nine Spits to escort us and the rest of the strike force—twelve Marauders from another South African squadron and Beaufighters from 252 and 603 squadrons (about twenty-seven Beaus in all—nine torpedo, nine rocket and nine 'suppression' Beaus armed with cannon). The Baltimores and Marauders were allotted the merchant ships.

It was about 7 p.m. and still daylight when we attacked. We managed to straddle a merchant ship and the South African Baltimores another. The others followed.

The rocket Beaus had a go at the merchant ships and some of the rockets went straight through without damaging them significantly. I think about six Beaus were lost but a South African box shot down a 109. Interestingly, the burning 109 passed right across my view through the bomb sight as we were on the run in and caused a bit of a distraction. There was a lot of flak from the destroyers but we were high enough to get away with it.

Of the three merchant vessels, four destroyers and six or seven MTBs [motor torpedo boats], only two merchant vessels, both on fire, and probably one destroyer made it into Candia Harbour.

So we were despatched early next morning to do them over in Candia Harbour. There were six flak guns looking after the harbour. This time we took a vic of six because unserviceability and damage from the day before kept the numbers down.

I remember going over Mount Ida—10 000 feet and snowcapped on occasions—into a very clear harbour.

We were at 14 000 feet and I picked out two vessels, one against the wharf and one out in the middle on fire. The aim was to drop a string between them to damage one or both. I think we damaged both with the bombs from the group of six.

Every aircraft was hit but none was lost. We were under fire for about five minutes on the run in and again on the run out. It was a bit distracting, but one of the things which relaxed me was my pilot. He was singing.

He was deaf but always insisted he was all right and wouldn't let anyone take him to the doctor. He wasn't too deaf for the intercom so I didn't make too much fuss about it. He was tone deaf as well and was singing a song with words to the effect that 'My tiny hands are frozen.' Somehow that made it much easier.

That turned out to be a very successful attack because subsequent intelligence indicated that all the merchant ships were sunk—one of them limping along on fire outside the harbour after our second strike—by a submarine. All the destroyers and escort vessels were sunk. This was on 2 June 1944.

Captured Ju-52.

It is significant that the Germans had come down to reinforce Crete and we had made that immense effort—immense for the Middle East at that time as we'd been denuded for Sicily and Italy. It was just before D-Day and fitted very nicely into the overall pattern. Reinforcements for France from that area were now not on because we were being nasty in that corner.

SICILY, 10 JULY TO 17 AUGUST 1943

After the American and British forces had driven the Axis from North Africa, the airfields of Tunisia were available for the invasion of Sicily. This began on 10 July 1943 and concluded thirty-eight days later.

Plans were then formulated to invade Italy. A logistic build-up was instituted in Sicily.

Jack Stronach took part in the transport support arrangements in September and experienced an unusual flying condition—wind shear:

I was flying a Dakota from Monastir in Tunisia to Catania in Sicily. I was crossing the plains round Catania with two engines strapped down in the aircraft and nineteen American airmen. They were going to a fighter drome in the Catania area.

I had come over some small mountains—about 5000 feet—on the south-east coast, and as I did so, I felt some updraughts. They got stronger and stronger.

I got to the strip where I was to land and was on the downwind leg of a copybook circuit. I lowered the undercarriage at the appropriate moment and I thought I was gone!

I hit a tremendous downdraft—straight down! These days it is known as wind shear—a narrow column of air going straight down at high speed. It can be very dangerous—particularly when coming into land, with the aircraft at low power and altitude with wheels and flaps down.

I thought the wings were coming off and were going to fold round the aircraft. This was a good demonstration of the need to wear a safety belt all the time. I wasn't wearing mine, and I later found that no one else was either. But I was the only one who did not hit his head on the top of the aircraft. I left the seat and was floating round in mid-air thinking "What the devil's going on?' The Yanks all hit their heads on the top of the aircraft—there was blood everywhere. They reckoned they'd get a Purple Heart out of it.

I eventually got the aircraft round and made a landing but, when I got out of the aircraft, my whole body was shaking.

THE BOMBING CAMPAIGN IN ITALY JULY 1943 TO JANUARY 1945

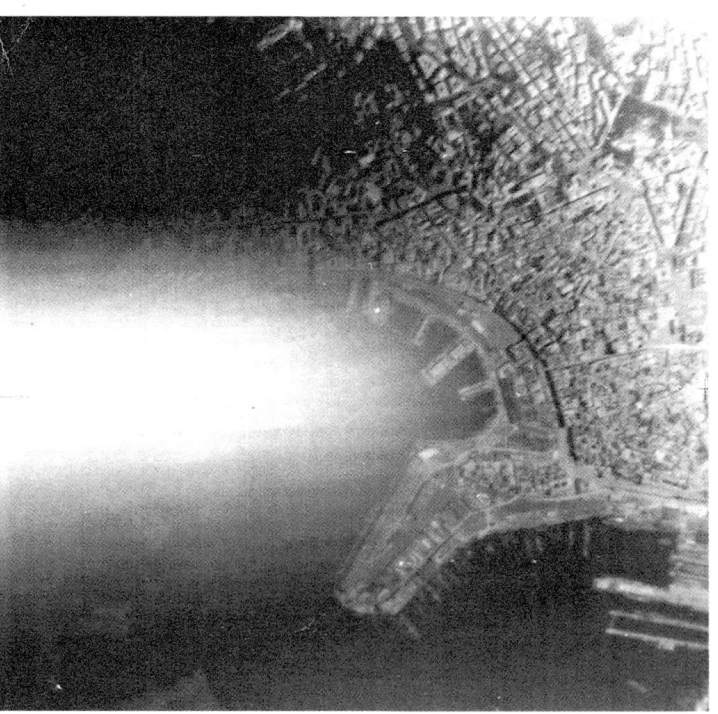

Target photo of Genoa 28/29th April, 1944.

On 3 December 1943 the XIII Corps, British Army, crossed the straits of Messina and landed in Italy. Five days later Italy sought and was granted, an armistice. However, the Germans fought on doggedly.

The British 1st Airborne Division occupied Taranto, at the heel of Italy, on 9 September. On the same day the Allies landed at Salerno, just south of Naples (the 'knee' of Italy).

Jack Stronach tells of his experience in the area:
On 14 September 1943 I took off early from our base at Catania in Sicily. I was flying an empty Dakota to Barcelona, on the north coast of Sicily. The operation was to move an American fighter squadron (Mustangs) to Sele in Italy.

My squadron took off, loaded with personnel and equipment, and flew in formation to the island of Stromboli where we were to rendezvous with the American squadron, in their Mustangs, who would escort us to Sele.

We flew to the coast of Italy, just south of Salerno. Then we flew down a valley. Next thing, we burst out and just in front of us was the armada, on the coast where the fighting was going on. We had to fly right to the other end. It was quite a few miles and quite mountainous country.

We reached the drome we were supposed to go to but there were red flares coming up everywhere. 'What's going on?'

Then I saw the balloon barrage above the Navy ships—hundreds of ships. At the same time we received a recall to not go in and the squadron started to turn away.

I thought, 'I'm going to have a closer look to see what's going on there.' So I broke and came right down on the deck and turned towards the coast. On the right, as I came in, was a tremendous wall of dust. It went right up in the air and, as I came across, there were these bloody Verey lights going off at me.

When I got back to Catania I found out what it was. It was the Germans coming back. Some of the fighters landed and were captured. the dust was the bloody shells bursting. No wonder I couldn't see anything. It was just like a big dust storm. The shells were bursting and the wind was taking the smoke and dust over the German troops. The Yanks nearly lost there.

By 15 September the 1st British Airborne Division patrolled in the area of Gioja and Bari. The airfield at Gioja was opened to the RAF. Six squadrons were based there and soon became operational. These aerodromes were supplemented by the two airfields at Foggia which were evacuated by the Germans on 25 September.

Bomber Command had launched its last raids against Italy on 24/25 July and thenceforward the Mediterranean Air Force had exclusive responsibility for RAF operations in this theatre.

Ron Turner, an RAAF squadron leader navigator, served with 37 Squadron RAF and tells how difficult flying conditions were:

Ron Turner with 'O' (observer wing) and crew members.

If you took off in winter you were covered with water—it looked like a flying boat taking off. As soon as you got airborne you had to open and close the bomb doors or they would freeze up and then, when you were over the target, you couldn't get rid of your bombs.

Ron tells of his experiences as a navigator:

I was in Wellingtons on my first tour and Liberators on my second. I flew in the Middle East—Tunisia—bombing mainly.

We bombed Italy and did some quite long trips. We had no sophisticated navigation aids and had to use our nous. I used a flame float to get the drift and if I really got into trouble I used the sextant.

I was lost only once, and that was because I was misled by the crew. The bomb aimer and pilot of the Wellington were sitting up front together and said there was a lake coming up and gave directions from it. I said, 'It's damn early! It's shocking weather and we should be late rather than early.' I did a dog-leg to kill time and ensure that I would not be early on the target.

But they had pinpointed the wrong lake so I didn't find the target we were to illuminate—this was in the days before the Pathfinders took over the marking of targets.

We were in real strife on the way home because I didn't know where we were and the weather was shocking. The pilot said, 'I could alter course.' 'What for?' I said. He said, 'I'm going down this valley.' This was all very well, but we were at 14 000 feet and in the middle of the Alps.

I got my sextant out and got a fix. Anyway, I got them home. The target was Styer in Austria.

Navigators could not see out—they were positioned in a little box. I used to come out, have a look at the target and then sit down in my office. If it was a long trip I sometimes went to sleep over the Mediterranean.

Jack Stronach, while serving with 178 Squadron RAF, took part in nuisance raids on Genoa:

I took part in four nuisance raids on Genoa. Our planes used to go over every hour on the hour. The effect was that the sirens would go, the workers would go down into the shelters, work was disrupted and it was hoped that morale would suffer also.

On 26 April 1944 I took part in one nuisance raid on Genoa.

We took off at 2200 hours and climbed to 11 000 feet to get over the cloud. It was very rough with many lightning flashes. After flying across Italy to just south of Naples, I set course for Corsica and ran into solid cloud and started picking up ice. It stayed like this most of the way to the target.

Just south of the target the weather cleared. It was a very dark night. The beam gunner started throwing out window to confuse the defences.

On the return the weather deteriorated. I descended to 10 000 feet to get away from the cold because I had lost the feeling in my hands and feet. The plane became iced up with ice flying off the props and hitting the sides of the aircraft. The clouds were full of electricity. The props were 'alight' with St Elmo's fire and the cockpit windows were lit up by sparks of electricity.

I had been flying on automatic pilot and had reached the Naples area. I turned on to a course to fly across Italy to our base when I thought we were gone. There was a blinding flash of lightning and the Liberator was thrown upwards a thousand feet in a matter of seconds. We had flown into a large cumulo-nimbus cloud.

When I realised we were still in one piece, I increased the power and climbed to over 16 000 feet and stayed above the weather.

On reaching base, I found it covered in cloud and flew around for some time before they put up the master searchlights over the city of Foggia to guide me down before dawn broke.

I started circling the cone at 10 000 feet, descended into the cloud and came out over the city at 9 500 feet in very heavy rain.

The navigator gave me a course to fly to the red beacon which was to guide us into base. When I left the searchlights it was pitch black and we were still in heavy rain. I didn't feel safe with the course I was on, particularly with the 3 500 foot mountain on the starboard side, into which two aircraft from my squadron had crashed.

I turned to port and flew back to Foggia where the navigator gave me a new course which brought us out over the red beacon. From there we flew to the flare path and went in and landed.

Dawn had broken by the time we got out of the aircraft and the mountain was completely obscured by heavy rain.

We were driven back for debriefing. It was there I found out why we had received such a hot reception over Genoa. We were the only aircraft that had reached the target. The others had returned because of the bad weather. Consequently, the window we had thrown out of the aircraft to confuse the defence had worked in reverse and alerted them to our exact position.

Nevertheless, only a few holes were found in the aircraft. That ended the twenty-sixth operational flight of my second tour.

Jack Stronach also tells of a very dicey moment before take off:

Sometime in May 1944 while on 178 RAF Squadron (Liberators), I had taxied our aircraft up to the end of the runway and had run through the cockpit check with the second pilot.

We were to be the first aircraft to take off and the rest of the squadron was lined up behind us. It was a pitch black night and all that could be seen was the lights of the flare path.

As the time for take-off approached, I taxied on to the runway and stopped, ready to go when we got the green light from the operations wagon.

Although I was now a captain on Liberators and carried a second pilot, I had developed the habit of doing a second cockpit check of vital settings myself. On the first check, with the second pilot, I had called for 10 deg. flap. When I checked the flap indicator on the second check the indicator showed full flap.

The mistake was soon corrected but had I tried to take off in a fully loaded Liberator with the flaps fully extended, I would have crashed. The aircraft would simply not have been able to become airborne.

I was not too pleased with the second pilot.

Jack Stronach tells of another trip to Genoa:

I had another unpleasant experience on an operation to Genoa. It was the last trip of my second tour, on 1 May 1944. I had climbed up to 18 000 feet but couldn't go any higher—I was too cold. My hands and feet had gone numb.

I started our bombing run, the bomb doors were open, but I'd never seen anything like the wall of flak that came up in front of us. It was a huge box barrage right on our height.

There was no way through that so I turned round, went out to sea, and came in again. I made three of these runs but each time I came in this bloody wall of solid flak came up.

So I said to the navigator, 'I'm going to come round and drop a couple of thousand feet.' I thought I'd be smarter than them.

Bomb doors were open and we were coming in beautifully—everything just about right—but they were right on to us. We were right in amongst the flak— shells were bursting all around us. I could hear them bursting and see the flashes and the black puffs of previously exploded shells were everywhere.

I said, 'We're over the target. Drop the bombs! Get rid of them!' The whole lot went down and I turned and put the aircraft into a steep dive. The gunners followed us down until we got out to sea.

Jack Stronach and his Liberator.

Sam Birtles operated in Baltimores from Italy. He was shot down in a raid on San Marchello. He relates how he felt:

On 23 August 1944 we were bombing in box formation, six kites in each, to destroy workers' huts near a little place called San Marchello. This was the fourth attack.

The squadron had been going back and forth for two days. We had been going morning, afternoon, morning, afternoon so the Jerries knew exactly when we were coming and where we were going.

We were on the afternoon shift and came up the Adriatic, out to sea to avoid the flak, and then in over the land from north to south.

The Baltimore was a very narrow-gutted plane and had a big perspex nose where I, as bomb aimer, sat. On the previous day we had twenty-six holes in us after we had bombed, so we knew that the flak puffs were a bit more than just smoke. On this trip I wore my parachute for the first time. Usually I just threw it on the floor— 'She'll be right mate.'

On this day we were with Michael (S/L Moore) as bombing leader. We were on his port side. When the leader opened his bomb doors I said to Keith Howard, my pilot, 'Open your bomb doors!', and when the leader started to drop his bombs I pressed the tit, counted the bombs falling away, and said, 'Close your bomb doors!'

I don't know whether he did or not. The very next thing was a mighty explosion which blew the perspex nose right off the kite.

My heart turned over. The next thing I knew I was hanging from my parachute on a pleasant summer afternoon. I saw two more parachutes come out. We had a crew of four. It was Dick Litchfield who was killed. He had replaced Jim McGrath who had a poisoned leg and didn't go on three trips. In each case, the plane he was going to fly in was shot down.

Chris Murray and Keith both got out and I saw the plane going down. It was on fire and exploded when it hit the ground. The petrol caught fire.

I was coming down gently and suddenly heard a crackling. I thought, 'Fire!' and looked up thinking my canopy was on fire. It wasn't. The crackling must have been ammunition exploding in the burning aircraft.

When I got closer to the ground—about 1200 to 1500 feet—I could see I was being blown backwards. So I took hold of the shrouds of the chute and tried to haul them round. It didn't make a damn bit of difference so I thought, 'Well I'm coming down, I'm not going up, so' . . . I kept my feet together and I put my bum prints into the soil.

ARMY SUPPORT ITALIAN CAMPAIGN SEPTEMBER/OCTOBER 1943–1944/45

After the Allied forces broke out of the Salerno beachhead in late September 1943, their progress was hampered by weather and stiff German resistance. Ultimately, Naples fell on 1 October.

The Allied forces then consolidated on a line stretching from the Volturno river in the west to Termoli on the east coast. There remained four major German defensive systems to overcome: (1) the Gustav Line, stretching across Italy from south of Gaeta to south of Ortona (including Monte Cassino); (2) the Caesar Line, from south of Ostia to north of Pescara; (3) the Gothic Line, from south of Spezia across the country to Fano; and (4) the Venetian Line, from Verona to Venice.

Gerry Judd was engaged in the defence of Naples and operations along the Volturno river area in September/ October 1943:

I was stationed at Fort de Leur, just outside Algiers. The aerodrome was called Maison Blanche. We were there, living in small canvas tents, for three weeks. That was when the Allied landings took place at Salerno.

When the Allies broke out of the beachhead and moved up to Naples, 600 Squadron left Malta and flew up to Pomigliano, just outside Naples. The Allied line was now right across Italy. It started at the mouth of the Volturno river and wound its way to Termoli.

At the time it was pouring rain. It did very little else but rain for the entire eighteen months. They had the heaviest rain they'd ever had. The rivers and brooks were all swollen. It was very difficult for an advancing army, as the Allies found out.

The Allies were busy clearing the port of Naples of the wreckage resulting from the German demolition work. A modern army runs on fuel supplies and so they needed to get fuel in as quickly as possible. The Germans were trying to mine the port and bomb the shipping in the harbour. Our job was to prevent them.

We had our work cut out. We had to stooge around at 15 000 to 20 000 feet so as to be on top of any raiders. But the Germans were pretty cunning. They'd come in very low over the hills at the back of Rome and then get right down on the water. They'd pull up over the harbour, drop their bombs, hit the deck again and get out.

We got some of them on the way out. The CO and one of the flight commanders each took out a Ju-88.

I was just flying patrols at that time—flying along the front line and waiting for something to happen.

Gerry Judd and his Radio/Navigator, Bill Brewer in front of their Beaufighter at Pomigliano, Italy.

That was my main war effort—constant patrols. You'd be two nights on, two off. You'd be allocated the dusk, the one after dusk or the dawn patrol. A patrol lasted four to four and a half hours.

Pomigliano was about four miles out of Naples, on the way to the Volturno river and Casserta, which was a well-known town north of Naples.

We were in tents again. It was always hard living wherever we went. 600 Squadron was a front-line squadron—never more than forty miles from the bomb [front] line. This meant moving quite often—just picking everything up, tents and all.

The American pioneer divisions would put a bulldozer through, perhaps a vineyard, put down PSB steel planking, and that would be our airstrip.

The weather was dreadful. Everybody was sloshing about. It wasn't too bad in the early stages. The rivers and streams ran down from the big mountains, where Monte Cassino was, into the Volturno and then into the ocean. But later, it became very muddy.

We were there when the attacks were made on the Abbey at Monte Cassino [part of the defences of the Gustav line].

We did the night defence above it and could see all the flashes. I think that the thing that impressed me most during the war, at least, close to the front, was the number of fires. There were always millions of fires.

Our patrol consisted of a run along the patrol line, orbit and come back along the line and repeat this for four hours or so. Then we'd come in, report and try to catch some sleep.

Some of Gerry's recollections of Pomigliano are:
Where we were under canvas we had a steel runway which was on the flatter part of the ground, while we were camped on the hillier part. The construction boys had run mesh wire around the camp because the Italians were all desperately hungry and driven to thieving. The camp was patrolled all the time as protection.

We used to go down to the flights in small trucks, crossing the main road between Naples and Casserta, which separated the aerodrome from the camp area. Where the road from the camp joined the main road was an old stone building where a couple of women had set up shop.

The ground crew had been out of England for a long long time and a number of them took advantage of the situation. Then a couple of them reported to the Medical Officer with some sort of disease. The Engineering Officer was concerned because these chaps were essential to keep the squadron flying. So he went down with a couple of his acolytes and a few packets of gelignite. They ushered the girls out and blew the place up. We had a party to celebrate.

A good friend of mine, a Canadian, used to billet together. We were right up against the mesh wire fence, away from the main body of the camp. The girls used to come up to the wire and call out, 'Hey Joe! Chocolata!'

The Canadian said 'These girls must have nits or something because all their hair is shaven off.' I said, 'That's not the reason. They were cohabiting with the Germans.'

Gerry encountered an 'odd bod' at this time:
We used to go foraging in the countryside for eggs and things like that. I tried to speak the local language and could make myself understood. We would hop into a jeep and scour the countryside bartering cigarettes or soap for eggs or chicken.

We had a fellow on the squadron who was a bit effeminate—or perhaps I should say he had a very effete or casual manner. His name was 'X'. He always wore long hair and had the propensity, when the truck was going down to the flights, to be the last on board. He'd say, 'I'll sit on somebody's knee.' Got a few looks of surprise.

The landing was made at Anzio while we were at Pomigliano. We were to be the night cover for the defence of the beachhead. We were briefed on the afternoon before the landing.

The briefing was very secret and took place in a big Nissen hut down at the flights. The door was shut and two guards posted. The Group Captain, who had commanded the squadron before I joined came striding down the hall and mounted the dais. He was a huge man, with a big booming voice.

Just as he was about to begin his speech there was a bang, bang, bang, bang on the door. It was opened—and in walked 'X'.

The Group Captain said, 'I might have known it was you 'X'. You long-haired dancing ponce.' 'X' said, 'I don't think that's at all fair. I can't dance.'

• Anzio

The Allied forces crossed the Volturno and proceeded up the Italian mainland as far as Monte Cassino, where they were confronted, on 15 January 1944, by the Gustav Line. There, the Germans were strongly entrenched and determined to hold on.

To draw off the German reserves, and threaten the rear of some of the German positions, it was planned to land on Anzio beach, in the rear of the Gustav Line.

The landings at Anzio, on 22 January 1944, took the Germans by surprise. The road to Rome was open. But the American general (Lucas) failed to capitalise on his initial advantage. In consequence, the Allies forces remained pinned down in the beachhead until 23 May 1944.

Gerry Judd:

We were given the night defence of Anzio while operating from Pomagliano. There was a Spitfire squadron stationed at Anzio. There was no aerodrome but the Americans just bulldozed buildings out of the way and laid a steel strip-runway. The Spitfire squadron was stationed at the northern end where they had a bit of a camp.

The CO said to me one day, 'Would you like to fly up to Anzio and take them a flare path?' The flare path consisted of just a line of naked bulbs that were half smoked on one side. Thus, you only saw them when you were coming in for your approach.

There were also some batteries etc., so they arranged for an electrician to get on board with me. They put all this length of runway flares in with me (wrapped up). I could only manage the load with the tanks half full.

We took off in my Beaufighter and when we arrived at Anzio I did a couple of circuits of the strip to get my bearings. Fortunately, you came in from the sea and the breeze was coming off the land, from the direction of Rome. I always used to make a steep approach and round it out so she sank on to the runway.

A friend of mine was on the ground and said later. 'We saw this Beaufighter orbiting around and said "Christ! He's not going to put in here is he?" Sure enough he did and the pilot got out and said, in an Australian drawl, "Where can I park this bloody thing?"'

My landing had shaken them a bit because a Beaufighter had a much longer landing run than a Spitfire. I had plunged in and eased back so she sank nicely on to the end of the runway, but she still ran on to the other end where they were camped.

After that they extended the runway by knocking over another old building which had been a motel. That made it not too bad to land. But landing there at night was still something. All you had were these bulbs which you could only see when you were on your final approach.

The runway lighting was like a T, with the bulbs along the side of the runway being the leg of the T. The top of the T was at the end of the runway. All you could do was swing right round and line yourself up. By watching the drift you found out which way the wind was blowing and then you'd come in and land.

We had two aircraft stationed at Anzio each night. They were on scrambles.

The Germans were coming in low over the hills at the back of Rome, swooping low and then pulling up to bomb.

Anzio Beach 1984.

Anzio was an IAZ [intense anti-aircraft zone] artillery zone so every gun was pointed up and was coned to about 12 000 feet to get these Ju-88s.

There were a couple of American cruisers in the harbour at the time and a number of other naval vessels. They had radar-equipped guns.

I was on scrambles one night and was given the 'go' to scramble. We'd been given information from Italians in the north that enemy aircraft had taken off. This was called the Y service.

I picked up a Ju-88 on radar about forty miles north of Anzio. We did an interception. I came in behind him but we'd come into the IAZ and he peeled off to the right. I attempted to follow and fire but the AA was trailing the German and took out my port engine.

In a Beaufighter all the hydraulics work off the port engine, so I gave an immediate call that I'd been hit. I dropped down to the deck to get away from the IAZ When you're flying through that stuff and the shells are bursting round you the one thing that sticks in your mind is the smell of the cordite.

I had to do an orbit to come in from the sea and the guns were still firing. I knew I was losing oil quickly. But the wheels wouldn't go down. They'd drop but wouldn't lock down. So I had to come in wheels-up.

Bringing in an aircraft wheels-up on steel planking meant that, once the propellers hit the runway, it was all chewed up. Also there were sparks galore churned out everywhere.

I told my navigator he would have to get out in a hurry. All he had to do was release his canopy catch and get out. 'As soon as we hit we'll get out!'

I had undone my harness already and, as we hit, I went forward and banged my head. But my helmet protected me. I jerked off my oxygen mask and hopped out.

I think I was out of the aircraft before it stopped. I had switched everything off and she didn't go up in flames. I was *lucky!*

I had to write a full report about it because I had buggered up their runway no end. They couldn't get an aircraft off the next day and the Americans were pretty browned off.

My view was that they were the blokes who had shot me down so it was there own bloody fault.

We were at Anzio for some time doing two nights on and then returning to Pomigliano.

The army were, in the meantime, slugging it up to Monte Cassino.

Churchill said 'We put a wild cat ashore at Anzio and it turned out to be a pussy cat.' because the American general refused to move. He was sacked.

Had they moved immediately, they would have done a pincer movement behind Monte Cassino and trapped the Germans in the mountains.

After the breakout from Anzio Gerry Judd moved up, following the army:
We moved up. The Americans had taken Rome, which was declared an open city. We went to a drome on the coast, just north of Rome, at Tarquinia. That was a pleasant place—we used to swim there.

There was a lot of conflict between the Americans and the British about how the war was being fought in Italy so the Americans elected to take the western side of Italy and the British and other forces took the east.

The Americans said they didn't want our night fighters over their area as they were going to put in their own people. So we upped from Tarquinia and flew across to Falconara on the other side.

Once again we were on the coast. The bulldozers pushed down the buildings and we had our runway. But, this time, we were lucky. All the Fascists had taken off and left beautiful homes which we used as the mess and our quarters.

• **The Gothic Line**

Kesselring, the German Supreme Commander in Italy, had to fight a defensive war as the Allies moved steadily up Italy. His defence line for the winter of 1944 was the Gothic Line. This stretched across Italy from a little north of Lucca, on the western side, to a little south of Rimini on the east.

1944 saw Gerry Judd patrolling in the vicinity of the Gothic Line:
A difficulty was the fact that the war in Italy was split between the forces of the various nations. The Americans were on the Mediterranean sector and we were on the Adriatic sector.

But the Americans found that their night fighter aircraft, the Black Widow, was a terrible aeroplane. The Americans told me it was a very heavy aircraft. Not only that, but, at that stage, they hadn't trained to fly at night.

What they got was only crash training programs. When they were coming in to land this aircraft they often wrote it off because of its problems. They finished up with scarcely any available for use. They then asked the British to supply night defence in their sector.

The blokes didn't like this at all because the Americans fired at anything. Flying up and down the bomb line meant you were just like those ducks in a shooting gallery—everyone has a shot at them. The Americans had all the ammunition in the world and just fired at any aircraft that wasn't one of theirs.

I used to fly on the German side of the lines and stay over there, well out of range of the American guns. This had the advantage that, if there was anything coming, I could get in early.

To get back to the squadron I'd keep on the German side until I could cross over to the British side. The poor bloody Germans didn't have any ammunition so you knew they weren't going to fire at you. Once on the British side I'd come over the Appenines, down to the plains at the bottom and back to the aerodrome that way. The day-fighter boys were at Jasi but the cloud was often so bad they couldn't take off.

Y service would somehow send a signal that the Germans had sent over some bombers. That meant we had to take off in daylight and climb through the cloud, which often went up to 25 000 feet. That in itself wasn't too bad because, if you were trained to fly at night, it was not much different flying in cloud. But it could be difficult landing if the cloud came down over the aerodrome.

The enemy did not pose us a problem in Italy. The difficulties lay in the weather and fire from our own side. We were being fired on all the time from the Allied side.

Cloud would often extend up to 25 000 feet and cause heavy icing. The icing was so severe we would get St Elmo's fire running along the wingtips. You'd swear the aircraft was on fire with crackling flames running along the leading edges of the wings.

The propellers would ice up and the ice would build up and slide off, hitting the fuselage. You'd swear you were being hit by flak. You were always fighting the weather.

The complete cloud cover didn't worry the Allied ground forces. In fact it made it easier to deliver their supplies to the front. The Germans were holding the Gothic Line very vigorously. They were not dropping back at all.

To drive them back, the Allies had to get their artillery up. But the roads up the Appenines were narrow. They were also icy. The trucks tried to get up at night but, on clear nights, the Germans would try to shell and bomb them. If the cloud came down, the trucks were just going off the roads into the ravines. In the end they decided to take the risk and drive up with their lights on. At least they could see where they were going.

The Germans then started to use Ju-87s at night. They'd come over and fly round until they saw a row of lights winding its way up the mountain. Then they'd dive and destroy the road and whatever else they could to block the road for a period. Of course, they couldn't do this in 10/10s cloud.

Our job was to prevent this. We would take off to patrol and have a call sign like China 2/3. You'd hear, 'Hello China 2/3! This is Gona (Base's call sign) we have a bogey for you. There could be a bandit. Vector one seven zero.' They'd try to bring you in behind it.

It was paramount with night fighters that you had to identify the aircraft before you shot it down. There were some dreadful mistakes made during the war.

The only aircraft which was difficult to identify was the Ju-88 because it was very like a Beaufighter, although the nose of the Ju-88 extended past the engines while that of the Beaufighter did not. A point not easy to check when attacking from behind. We did know, though, that we were the only Beaufighters flying in the area.

When you got up behind the 87s you could identify them by their big cranked wing. They were a big aircraft and the wheels hung down. They would be weaving all the time until they sighted a target, when they would straighten, put the nose down and go into a vertical dive. This was all at night.

My operator would say, 'I've got a contact at (say) 2000 feet at 2 o'clock. Closing.' I'd tell him I'd got it when I could see the silhouette weaving in front of me. As soon as I had identified it I would get underneath it. He had to straighten up to drop his bombs and as he did, I'd come up behind him and give him a burst.

The Beaufighter had the greatest firepower of any fighter aircraft. You had four 20 mm cannon and six .303 machine guns. I always tried to get in close to make sure of getting a kill the first time. This did cause a problem once though. The fourth one I got blew up and I lost my port engine and my hydraulics.

I had to limp back on one engine. I went all right until I came in to land. I didn't use the port engine at all. Then I gave it a kick and the wheels came down and I got down safely.

I had my father and my brother in the army. My job was to shoot down German aircraft. I had no animosity towards soldiers.

One day I was on a dawn patrol on the German side of the lines. Coming back, I flew over the Appenines. Daylight was coming through and it was quite light. I was losing height, from 20 000 feet to about 15 000 feet, and just coming over the crags. The mountains in Italy are not like the mountains in Australia. Kosciusko is a saddle and you gradually go up. Over there they were crags going up to 15 000 feet and had big valleys.

I'd eased the throttle back and was letting the aircraft sink—maintaining speed but just steadily losing height. My radio observer was occupied with his log.

I indulged myself with some low flying, which every

pilot gets a kick out of—particularly after four and a half hours of patrolling. I came down a ravine and into a wide valley with a river running through it. On both sides of the river there were hundreds of fires and tents. At the end of the valley I got down quite low—say 100 to 200 feet—where there were some horses. As I got down there I could see the horses had big ears. Then I suddenly realised they were not horses but mules. I lifted the aircraft up over them and dropped down the other side.

Only then I realised that the troops were Germans. A brigade or half a division was camped up there.

When I got back the Intelligence Officer (he was an Englishman who had lost an arm when he was flying) said, 'Anything happen?'. 'No. Pretty straightforward.' And then I told him what I had seen. I didn't think any more about it until the next day. The CO said there was a major there from the army.

So our bloke and I went to a tent and he asked where the encampment might be. He asked me to show the position on the map. I said, 'It's pretty difficult. I wasn't taking much notice and my observer had his head down. I was relaxing.' He said, 'Surely you have some idea.'

He said,' Make a stab' and I stabbed wide—near the American sector. I knew that wasn't right but I thought, 'If I say exactly they'll bomb the poor bastards and they weren't doing any harm—just camped up there.' The wrong thing but I felt good. I felt sorry for them, camped in the freezing conditions and merely being defensive.

Peter Matthews, of 454 RAAF Squadron, flew Baltimore bombers in this phase of the campaign. He flew from Falconara in support of the British 8th Army:
I joined the squadron in October 1944. It was at a place called Falconara, which is on the east coast—about half-way up. We were support to the 8th Army and weren't far behind the main lines. The squadron personnel could actually see the anti-aircraft fire directed at us and could sometimes hear the detonations from the front line. These were not necessarily from our bombs but from the shelling by both sides.

On some occasions we were very close to the front and, when the Germans moved north, we did also. When we moved from Falconara to Cesentico the front line was only about sixteen or twenty miles away.

We were usually engaged in close support raids. We would fly in three vics of three. The squadron would join up with other squadrons en route to the target.

The bombing leader was in the key navigational plane. He controlled the bombing. He would open his plane's bomb doors and fly straight and level for thirty seconds. Then he would release his bombs.

All following aircraft dropped their bombs at the same moment as the leader, thus giving a saturation coverage of the target. If the leader had made a mistake, then, hopefully, the vic behind him would correct the error.

Nevertheless, on these close-support trips, all the crews were fully briefed and given the met information, height to fly at, etc., so that, in the event of being separated, the navigator could get us home.

THE BALKANS
14 APRIL TO 18 AUGUST 1944

The Mediterranean Allied Strategic Air Force had, during this period in particular, the task of attacking the enemy in south east Europe and the Balkans in order to support the Soviet land operations. The main components of the force were the 15th US Air Force and 205 Group RAF.

This section focuses mainly on 205 Group's role. The Group consisted of four wings: 231 Wing—Squadrons 37 and 70; 236 Wing—Squadrons 40 and 104; 330 Wing—Squadrons 140 and 150 (which operated with Wellington X aircraft until the second half of 1944); and 240 Wing—SAAF Squadrons 178 and 31 (which operated Liberators H or J).

Over this time 205 Group mounted more than 1000 sorties against targets such as the Rumanian oil industry and the road and rail networks supporting that industry.

The Danube, a major waterway, was bombed and mined—mining was the major activity. About 1200 tons of explosive were dropped and over 280 mines laid.

The defences encountered were generally not as heavy as those met by Bomber Command in Europe as the latter had to deal with a much greater weight of attack. Nonetheless, the USAAF Intelligence claimed that the defences of the Bucharest / Ploesti area were only surpassed in defensive strength by those of Berlin and the Ruhr.

• Hungary and Rumania

Mining operations were planned jointly with naval staff, who advised on the type of mine to be used and the nights when mining was most likely to be effective.

The level of the water in the river was an important factor governing the height at which the mines should be released. In summer the water level was usually low and the release height was then about 200 feet.

Jack Stronach tells of his minelaying experience:
On minelaying operations we used to fly across the mountains of Yugoslavia, which were about 10 000 feet high. We'd go over at about 12–13 000 feet and drop down to about 1000 feet until we found the river. Then we'd come down to 150–200 feet — sometimes down to

Jack Stronach and crew, Celone near Foggia, Italy. *Back left to right:* Jack Stronach (1st pilot), 'Mac' McNaughton (2nd Pilot), Doug Howes (Navigator), A.Sproates (Wireless/Air gunner) *Front:* W.Plowman (Air Gunner), Geordie Temple (Air Gunner), A.Burrington (Air Gunner).

fifty feet. It was hard to judge your height. We were given a barometric reading to set the altimeter before we left base. Essentially, you had to go by your own judgment.

We had to do a dummy run before dropping a mine. They were very secret types of mines. They had a parachute at one end of them. We carried six in the Liberators and two in the Wellingtons.

Our aircraft was cramped. I remember trying to get in. You couldn't get in with the parachute harness on. You had to get in and then put the harness on inside. Just as well I was nice and slim in those days.

You'd do the dummy run and then peel off to port and come in to do the drop. Of course there were a lot of AA guns and they'd slam a few at you as you went past. I remember on one trip we could see the flak coming up ahead of us. It was being fired at the Liberator ahead of us. I told the gunners we would do a turn and to get ready. The first tracer to come at us would mean a burst at the flak gunners. But they didn't fire at us.

I was looking for any movement of shipping. I came to a bridge across the river—it was a moonlight night and I told the beam gunners to put a burst into it. The tracer hit the bridge. Now they knew we were coming and they poured it back at us. Their tracer came very close. On returning to base we found the Liberator ahead of us on the river had been hit and the rear gunner had been killed.

Ron Turner, a squadron leader navigator, served in the Middle East command and has some random observations about attacks on Hungary and Rumania:
We bombed Bucharest from 11 000 feet, Budapest from 9000 and Plodiu from 6200.

Over Budapest we collected the most holes we ever collected—some sixty-two hits. The weather was bad and we got knocked around over the target. There were sixty or seventy kites on the raid. The searchlights were just like the sun. However, I sat in my office and I'd only come out to have a look when we did something silly like diving round the sky. You had to do this to get out of the searchlights because they'd flick on, and if the master light was on you the flak would come bursting round you.

We went in to every target on 'George' [automatic pilot] and that gave us a nice even bombing run.

We flew to the Ploesti oilfields from Italy—about an eight-hour trip. We weren't attacked ourselves by fighters but a hell of a lot of blokes were.

The defenders became used to us passing over and you'd be flying along most peacefully when, all of a sudden, you'd be hit by shells from a mobile flak gun.

On one trip to Budapest we were thumped all over the sky. I think we were one of only four planes out of the two squadrons which reached the target so they had us pretty much to themselves.

Mining sorties in Wellingtons in the Middle East were quite different to those in the European area, as evidenced by Ron's experience:
We were detailed for mining operations—to mine the Danube at Budapest. We had to go in low—fifty feet.

Our tactics caused our rear gunner to almost have hysterics. It was like skip bombing. The mine dropped from the centre of the aircraft and the bounce nearly took him out of his turret. He was screaming his head off.

A close shave as related by Ron Turner:
We mainly flew across the sea en route to our targets. On one occasion the Jerries got on to us, so Max put the kite into a steep dive.

The poor old rear gunner kept screaming as they followed us down quite nicely. But they just missed us as we got down on the deck. The gunner nearly had a heart attack. He reckoned he could have very nearly reached out and touched our attacker.

(The rear gunner, be it remembered, sat in a turret well away from, and out of the sight of the rest of the crew. He faced long hours of isolation and darkness, cramped in a small cold space. He had to stay alert, constantly searching the night for enemy fighters. Little wonder he was 'a bit of a screamer.')

• **Yugoslavia**

Yugoslavia was also attacked and the naval base at Pola, on the Adriatic Sea, was the focus of much attention. It was a target which crews approached with a great deal of apprehension.

Peter Matthews, a Nav/B with 454 RAAF Squadron, had this to say about a sortie to Pola:
It was my first raid and we carried it out on Pola in Yugoslavia. It was a very heavily defended naval base. When our crew first got to the squadron all the existing crews were talking about targets and one of the names which kept coming up was this name—Pola. Nobody wanted to go there because it was the scene of great anti-aircraft activity and, therefore, to be avoided at all costs.

You can imagine our feelings. Our first raid and the first time we'd seen ack-ack. We saw it! Black clouds—some of our blokes reckoned they could smell the cordite. All hell broke loose because of the tremendous fire. We all took independent evasive action. This meant diving, breaking our normal vic and getting to hell out of the place. We were thrown about a bit and headed towards the sea.

But, in all this violent evasive action, our pilot inadvertently knocked a feathering switch[2] with his elbow. This cut out one engine and we began to lose height rapidly.

The pilot didn't know what was wrong and thought it must have been the flack. We were all considering what was the best thing to do. I'll admit there was some confusion. After what seemed an eternity to me, the pilot realised what had happened and managed to get the engine back to life.

George Gray provides further comment:
At 10.10 a.m. on 2 December 1944 Squadron Leader Cashmore and myself took twelve Baltimores, with an escort of five Spitfires, to have a go at warehouses and a possible ship in Pola harbour.

The anti-aircraft fire was very heavy and most of our aircraft were hit on the run in. One or two aircraft skidded and their bombs fell on an infantry barracks that were near the wharf.

I think we were one of the first to get our bombs away but, just as we were making our run in, all hell had broken loose and the flak included some 'flaming onions'—an anti-aircraft rocket (fascinating as it came along the drift wires of the bomb sight) which was not very accurate. Suddenly I was not in communication with my pilot.

In a Baltimore, the nav/bomb aimer is in an isolated compartment in the nose and the pilot was up above—you could see where his feet were. In case the intercom broke down—as it had now—the Nav had an air speed indicator and a control column, which he could swing out, and a throttle. I used to use this on low-level sea reconnaissance to give the skipper a rest.

On this run the intercom had gone out and the pilot thought his new bomb aimer had 'had it.' He thought he'd make the run on his own but found that the bombs had gone.

Due to the flak there were aircraft all over the place. We finally managed to get them together into some sort of formation. It was pretty ragged but the attack had been a success.

The other thing I remember, apart from our concern, was going out in an easterly direction, with the fighters around us and going in tight. The sunlight was beautiful coming off the snow-capped mountains behind Pola. It was like a holiday trip until we suddenly realised that the game was on.

Quite a number of crews had to ditch on this operation and the Air Sea Rescue people were pretty busy picking them up. Pola was a rough target in anyone's book.

The Air Sea Rescue went out, under naval and air cover, to points half-way across the Adriatic so they'd be on hand every fifty miles. They mainly used motor torpedo boats. They'd turn up at high speed, so we knew we had a chance. If we made a straight line for home we knew there would be someone there who could look after us.

---

2. Feathering switches were provided to allow the pilot to cut an engine in the event of damage. The feathering switch stopped the propeller—and hence the engine—rotating. This avoided the risk of the engine disintegrating and, more importantly, reduced the drag of the otherwise rotating propeller and improved the chances of getting the damaged aircraft home.

THE WARSAW UPRISING

On 25 July 1944, the First Byelorussian Army, under Marshal Rokossowsky, reached the Vistula. The Germans commenced to retire in anticipation of an early Soviet attack.

The Poles, whose country had been overrun by the Germans in 1939, had instituted a system of resistance groups; some active, and some passive but ready to act when the time was opportune.

The leader of the Polish resistance forces, General Bor-Komorowski, had about 30 000 men and women in his force—mostly resident in the city of Warsaw. Judging the time was right for his force to rise up and attack the Germans, he ordered the attack despite the fact that he had no heavy weapons and only a limited supply of automatic weapons. This might have sufficed for a short sharp action.

But the Russians paused and did not attack. The Poles were deserted. The Russians would not even allow Western aircraft, engaged in dropping supplies, to land on Soviet airfields. Only on 12 September, when it was too late, did Stalin agree to allow it.

The Western Allies did what they could to supply the gallant Poles. Polish, British, South African, New Zealand and Australian airmen flew supply missions from Brindisi in Italy. They were later joined by Americans.

The flight involved a 1700-mile return flight. Losses were high and successful drops few.

Ron McCathie, a squadron leader with 454 Squadron, had personal experience with American airmen—as he relates:

454 landing strips were always within ten miles of the front line. They were bulldozed through vineyards, orchards or open fields and then clad with interlocking metal strips. They were not very substantial, but adequate for damaged friendly aircraft coming in from the north.

The people of Warsaw had risen en masse and attacked the German garrison troops. These courageous people finished up occupying half the city and, in the process, pinned down two German divisions for about three months.

During these months, supplies of munitions, food and medicines were dropped in to them by the United States Air Forces based in southern Italy.

Some of the other Allied aircraft used in these supply drops had Australian navigators. The round journey to Warsaw took twelve hours—most of the time in daylight.

The Americans only had top fighter cover for a third of the way up and a third of the way back to Italy owing to the limited fuel capacity of the fighters—which were mainly Mustangs. They flew north in tight formation and were able to defend themselves. But coming home after a casualty-fraught low level drop over Warsaw was a different matter. They were picked off, one by one, as they limped back to their home base. Casualties averaged 50 per cent.

On the days these Warsaw supply drops occurred our strip became a shambles. 'Lame ducks' came in every ten minutes; completely cluttering up the strip. Our operational program was ruined.

These USAAF blokes were flying Liberators and Fortresses. On one of these days a Liberator came in and landed. Only two motors were working—and one of those was coughing.

Fortunately, the hydraulics must have been OK because the wheels came down. Our people on 454 waited to see the crew scramble out when it stopped gumming up the runway. Just one lonely little man climbed out of that large aircraft.

It transpired that, on reaching the Adriatic Sea, and prior to making his landing approach with two doubtful motors, he told his crew to bale out. They did and were all picked up by the Air Sea Rescue within the hour.

That very nice little Texan, twenty-two years of age, stayed with us in the mess for two weeks while the salvage people got his aircraft serviceable. It had been his fourth mission to Warsaw. The tales he told would make your hair stand on end. He was living on borrowed time. Those Yanks were very brave men.

This experience left us more than happy with our lot.

It was a great pity that the people of Warsaw were eventually overwelmed despite our efforts to support them.

In January/February 1945, 454 Squadron concentrated on intruder bombing by night. They generally sought to make the German retreat as difficult as possible.

Peter Matthews took part in this effort:

In January 1945 I started to convert to night-intruder duties. In these activities we were virtually given an area in which we could do whatever we wanted. We could fire at whatever targets we could see, or photograph bridges, or bomb troop concentrations or bridges. We didn't go out to bomb any specific targets but attacked 'targets of opportunity'.

We'd operate at 200 feet—at night! Italy was crossed by a lot of electric transmission lines going in all directions and they were what we had to be most careful of.

Most of our night-intruder operations were in the Po Valley which was pretty flat; so it wasn't quite so hair-raising as it seems at first sight.

I had an unusual incident on one mission in the vicinity of a place called Udine—in northern Italy, roughly between Venice and Trieste, at the foot of the Alps.

We were attacking barges and bridges and, all of a sudden, became aware of a red light. Wherever we went this red light went and we broke radio silence to report it.

There were no other friendly aircraft in the vicinity—or shouldn't have been. We took gentle evasive action—diving and doing 180 deg. turns—but the red light stuck like glue. We just couldn't get rid of it.

The rear gunner opened fire on it—he was a bit of a gung-ho type. The skipper said it was most unwise to aggravate it, increased speed and broke off the engagement. After a few minutes we lost it.

By coincidence, when the war finished, our base was at a drome called Tempo Formile, near Udine. We were told by the Italians that a night-fighter training unit had been operating there.

We concluded that our red light had been a chap on a training flight and he wasn't too sure what he was doing. We weren't sure either. Why he didn't knock us over I'll never know as we would have been a sitting duck.

Crashed Baltimore El Djen, Tunisia.

3

SOUTH-EAST ASIA

Laurie Jones' Liberator.

On 4 March 1944, Autralian Prime Minister Curtin sought reassurance from Churchill regarding Japanese naval moves.

Churchill replied 'Possibly the Japanese intend to raid our communications between Calcutta and Ceylon, or elsewhere, in the Indian Ocean . . . Our battleship squadron in Ceylon is well posted. Our shore-based aircraft are strong.'

Laurie Jones was a Liberator pilot who completed a tour of operations in Ceylon. The story of his tour gives some idea of the war fought from Ceylon:

I joined the squadron, a Coastal Command squadron, at Sigiriya in Ceylon. Sigiriya has a great 600 foot rock which is of great religious significance to the locals. We didn't appreciate it much because it was right in the middle of our circuit. It had been hacked out of the jungle at short notice under threat of attack by the Japanese fleet.

It was not really suitable for Liberators. The strip was relatively short and rough with a hump in the middle. If there was a Liberator at the other end of the strip you couldn't see it. The jungle curved in over the edge in an attempt to make it less obvious to enemy aircraft but it was lethal for heavily laden Liberators taking off at night and we lost a number of crews.

Our initial role was convoy escort and anti-submarine work.

Just as I arrived two crews who had specialised in photographic reconnaissance were shot down over the Andaman Islands in the Bay of Bengal. A Canadian Flying Officer and myself volunteered to form a new photo reconnaissance unit. We started operational work after some self-training.

We had two roles. One was mapping Sumatra and islands in the vicinity with the K17 Survey cameras. The other was to record Japanese activity in ports and other installations with the F52 reconnaissance camera. We also joined the rest of the squadron in its convoy escort and anti-submarine activity.

The weather in our operational area was generally appallingly bad. We had often to wait nearly two weeks before it was worth crossing the Bay of Bengal. The trips were about fifteen hours duration.

To get to the neighbourhood of Sumatra we had to take off at night to get to the area at first reasonable photographic light. We had to fly through the Inter Tropic Front over the Bay of Bengal and that was a frightening exercise.

Clouds extended from virtually sea level to 40 000 feet and the updraughts and torrential rain were almost mind boggling. We had to go through the front because it extended for hundreds of miles across the Bay of Bengal. The alternative was to go underneath. I tried on a couple of occasions.

It was at night, of course, and trying to sneak between the base of the thunderstorms at 500 feet and the sea was an exciting experience. The noise if the rain was teeming down and hitting the aircraft was almost bewildering and it was almost impossible to see what was sea and what was cloud. We were more like a submarine than an aircraft.

We carried two pilots, both had been trained as navigators. Coastal Command pilots had mostly done a General Reconnaissance (GR) course which was basically a navigation course. The rest of the crew were the navigator, wireless operator, flight engineer and gunners.

The Liberator was a fairly comfortable plane and we had sufficient room to take a reasonable amount of food on board. We'd normally take off about midnight and get to the target about first photographic light, spend about half an hour in the target area and then relax and settle down to a very substantial meal. I notice in my diary I used to demolish a tin of Spam, tin of peaches, fruit cake and another three or four major items— probably just the relief of finishing another operation.

I would normally go into the mid-upper turret and doze for an hour before making our landfall in Ceylon. We were operating up to a thousand miles from our base with no escort. A solitary Liberator was no match for the Japanese Tojo fighters. If they found you it was largely a matter of luck whether you could get into a cloud and get away. Otherwise your chance of survival was remote. I always managed to evade.

I was once over the west coast of Sumatra. I had gone over to do some more survey work but the middle of the island was clouded over. We did a leisurely run up and down the coast with the cameras rolling. We didn't hurry because the main Japanese base at Medan was on the other side of the island and as it was clouded in we assumed there wouldn't be any trouble.

When the photos were developed we found there were six Tojos in the act of taking off from the satellite strip. They were just below us and as we were the only aircraft in the area they were about to pursue us.

When we did convoy work we normally had two or three aircraft patrolling round the convoy. Generally the presence of patrol aircraft was enough to ward off submarine attack.

We had eight or ten depth charges for anti-submarine work. The aim was to drop a stick at an angle of 45 degrees to the axis of the submarine. That gave a good chance of getting two (one on either side) charges in the right position and that should be enough to kill it.

We had very few attacks. The fact that we were in the area was enough to keep the subs away from the Ceylon–Australia shipping route.

We did, however, have a very unpleasant U-boat in our area. We nicknamed him 'The Butcher' because he left very few survivors. We hunted him round the Indian Ocean but I don't think we nailed him.

I was only basically concerned with submarines and photographic reconnaissance. However the rest the squadron had developed long-range cruising techniques. They lightened the aircraft and used improved fuel which meant they were able to remain airborne about twenty-two hours.

They conceived the idea of laying mines in Singapore Harbour. They sent eight planes on the first night. Across the Bay of Bengal, around the northern tip of Sumatra, down the Malacca Straits and into Singapore. Seven of the eight made it there, dropped their mines and reached base safely.

The squadron kept going back. We were probably doing the longest strikes of any in the RAF at the time. Twenty-two hours was the shortest but they sometimes got close to twenty-four hours.

There were no radio aids and the long hours became particularly gruelling for navigators. But we were carrying, in effect, three navigators.

SOUTH-EAST ASIAN THEATRE WWII

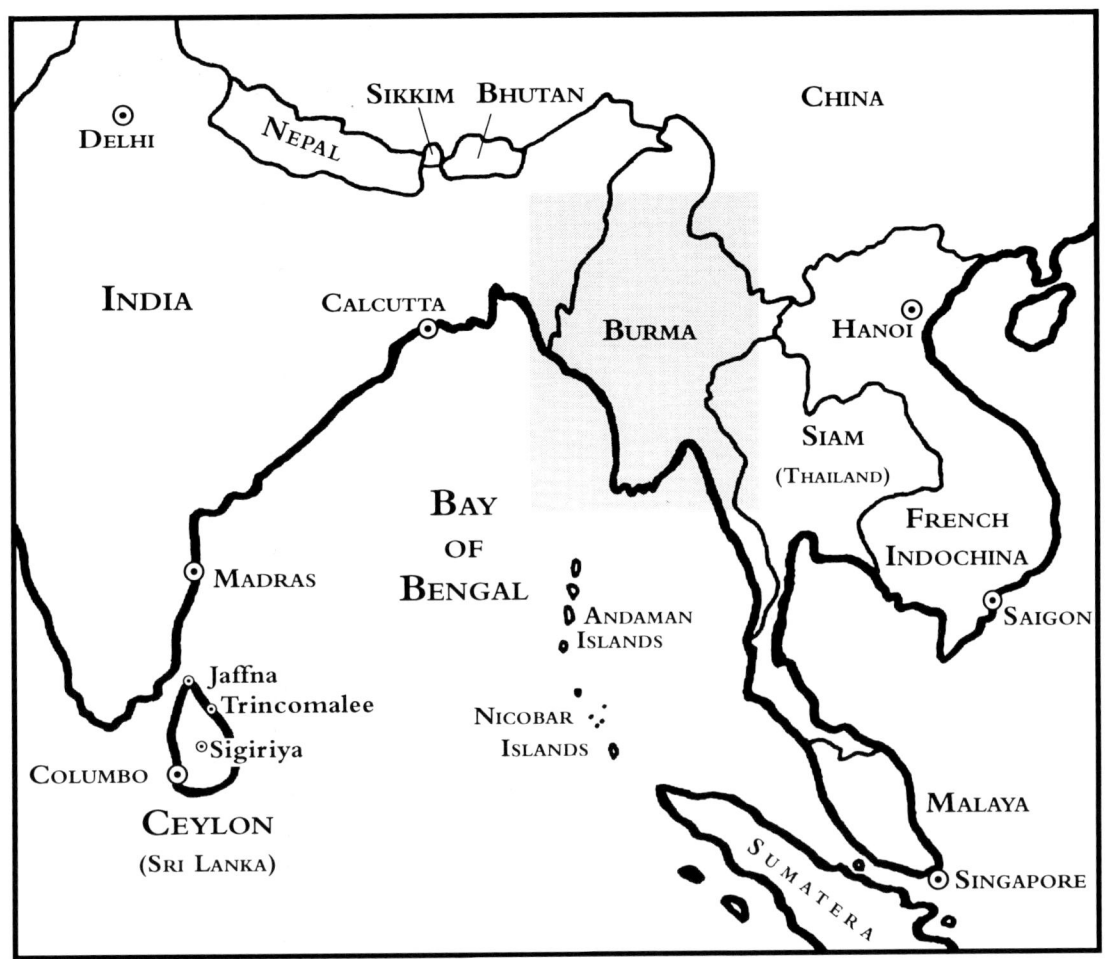

At the time I was about to leave the squadron (about May 1945) we started dropping people and supplies into southern Thailand and Malaysia. We were operating a long way further south than most people in the Burma theatre.

I had to do a low-level reconnaissance of Phuket Island on the Kra Isthmus in Thailand. It was possibly the most important trip of my career. The aim was to photograph the beaches in preparation for a landing which was intended to capture that part of Thailand and then move south into Malaysia. My trip was planned for some time and PR Mosquitoes flew high-level photographic reconnaissance for us to use in planning our sortie. Their photos showed heavy ack-ack and fighters in the area.

We flew across the Bay at about 5000 feet and dropped to a couple of hundred feet as we approached the gap between the Andaman Islands and the Nicobar Islands. We kept that level all the way to Phuket.

We flew over Phuket at 1000 feet to get the precise coverage that the army was requesting, high-water mark, low-water mark, any obstructions, shallow water and the positions of the gun emplacements round the invasion beaches.

The pictures from this trip were deemed so important that a submarine was diverted from its normal patrol to the target area in case we were shot down or disabled. If we ditched near the submarine it could ensure the rescue of the film and, incidentally, the crew.

Flying for lengthy periods at low level requires hard concentration. It's not too bad at 200 feet but it's a bit bumpy. In General Reconnaissance squadrons you spent a lot of time learning to fly at an estimated fifty feet which is the height at which you drop your depth charges. Your low flying needed to be accurate and long practice was necessary.

They built another aerodrome for us and we moved to Kankesanturai in the north of the island, up towards Jaffna. This was a fine strip, big, wide and open but it was further away from our objectives on the other side of the Bay of Bengal. We had to move a lot of our PR work to China Bay which was the big naval base at Trincomalee.

David Corthorn also gravitated to KKS as he calls it and was flying Liberators:
En route to Ceylon, we had to get repairs and maintenance done on our aircraft at Cawnpore. Instead of taking three days it took a month, due to the number of religious holidays to be observed. Out of a seven-day working week you were lucky if you got three days, sometimes only two. We were able to spend some time on leave.

Then we went to the interior of Ceylon, to a place called Sigiriya. It was very high up. Our next station was in the north of Ceylon. From there we operated to Trincomalee, and across the Bay of Bengal to Sumatra. We were patrolling—mostly looking for enemy shipping or ditched aircraft.

On 24 July 1945 we got a message that a Liberator had ditched off the northern tip of Sumatra. We took off—the trip lasted thirteen and a half hours. It was an interesting trip, because I had to do astro-navigation [navigation by using a sextant to measure the altitude of stars]. We didn't find any evidence of wreckage, or any survivors, when we got there, so we flew back.

After the war, I went to live in Lane Cove. I met a lady who had lost her brother off Sumatra. She could never find out if anyone had looked for him. I had! She was very relieved to know this.

From our base in Ceylon, we did patrols in Liberators, back to the Persian Gulf and the Arabian Sea.

Laurie Jones takes up his story again:
At China Bay we were joined by two Mosquito PR crews from 604 Squadron and we operated together for some months. The Mosquitoes did the closer targets and the Liberators did the longer. The 604 boys had a Mitchell which was used to survey Southern India where the maps were poor. The PR aircraft were pale blue and the others were white underneath and normal green camouflage on top. All our photography was from 25 000 feet.

There was to be a large-scale invasion of Sumatra. The troops would then move to Malaysia and Singapore. I spent my time making maps of Sumatra.

The normal sortie involved flying across the Bay of Bengal, through the Inter Tropic Front at about 5000 feet, then climb to 25 000 feet about an hour before the target. After about an hour and a half we'd come down quickly to 1000 feet and head for home. There wasn't much flak on these trips.

There was several changes of clothing on these trips. The jungle airstrip was incredibly hot and humid. I used to take off in my underpants and sitting on a towel with the sweat pouring off me. When we were at 5 000 feet across the Bay we donned normal shorts and shirts. As we started to climb we put on battledress, sweaters and sheepskin clothing and boots. The heating was pretty good in the Liberator but you still needed heavy gear. Then, on the way back we started taking it all off again.

The plane was always a shambles of clothing, parachute packs, survival equipment, machine-guns, kukris and revolver belts. It was almost the fashion for South-East Asian crews to carry a kukri. If it was good enough for the Gurkhas it was good enough for us. We'd have a .38 Smith & Wesson on one side and a kukri on the other. We made sure the kukris got nowhere near the compasses. By the time we got ourselves into the plane it was shambolic.

The Americans mounted a B-29 raid from China Bay. As they returned they put out a distress call. The 604 boys went up in the Mitchell in appalling weather and bought it while the B-29 landed safely.

There was a large strip at Minneriya in south-eastern Ceylon and a number of Liberator squadrons assembled there in late 1944 and 1945 when things quietened down in Burma. We became one of a number of resident squadrons instead of being alone on a base.

It was a huge improvement after our early days at Sigiriya. The jungle was not like that at Sigiriya with its monkeys and other inhabitants being with you and all around you. There we had had pretty basic accommodation—jungle-huts made of palm leaves and so on. Minneriya was most civilised by comparison.

It was about this time that Qantas started to operate again and the highlight of our existence was watching the Lancastrian taking off during the middle of the day. It would roar down the strip and disappear over the hump in the centre. Then we'd see a small speck fighting for height in the distance.

I was sent back to India to get an all singing and all dancing photo reconnaissance plane. Our existing aircraft were fitted with four cameras—three F52 reconnaissance cameras and one K17 survey camera. This was pretty primitive, as were the heating arrangements for the cameras. We had all sorts of muffs to try and keep the cameras warm and operational but they tended to freeze up at 25 000 feet. We sometimes returned home after a sixteen-hour trip to find the cameras had frozen up.

It was decided we should have something better so a Liberator was commissioned to be converted for photographic reconnaissance work. I went up to Cawnpore to pick up this aircraft. It had two K17s, three F52s and two K8ABs which were the latest and had a marvellous camera bay which was properly heated and had all the refinements we hadn't had.

I flew it down to Calcutta. It worked superbly for a time but two engines gave out and we had to take it back for new engines.

I was now tour-expired and handed it over to a new crew who were to fly it to Cocos Island and operate over Java from there. They loaded on the full operating crew, the ground crew, the PR intelligence unit and our army mapping man—the core of our PR unit. It took off from Minneriya, lost an engine and crashed losing the whole nucleus of our PR unit.

On the ground, the aircraft were like furnaces inside. The mechanics working on them often passed out with heat exhaustion. If possible you avoided flying in the middle of the day.

Looking at my log book I find we had a fair amount of unserviceability but most of it was at Sirigiya where the planes were dispersed in the jungle. It rained a lot and it was incredibly humid—the worst conditions for sophisticated electronics. Some of the Liberator's sophisticated equipment didn't fare as well as the more basic gear in some of the British planes. Our ground crews were incredibly good—they worked their hearts out.

We only had one crew ditch and survive. They were mining Singapore. It was one of the Canadian flight commanders—one of those who had developed our long-range cruising techniques.

Off the east coast of Ceylon they ran out of fuel after twenty-four hours. Their Mayday call was received and a rescue craft picked them up but the squadron leader and two other members of the crew were taken by sharks. That was the only time anyone crashed and got back.

I finished in May/June 1945. I had crewed up in Nassau and had five Australians in the crew. I was the only Australian captain so any Australians about were keen to join me. I started in December 1943 so I had about eighteen months on the squadron.

INDIA

Conditions for flying in India were very different from those in Europe. White outs, where visibility was nil, occurred over the subcontinent when monsoons swept across the country. Doug Nicol gives a picture of the conditions experienced while operating there in 1945:

I remember a trip out of Karachi, when we flew in the vicinity of the Eastern and Western Ghats. The latter are inland from Bombay.

We were in the air when the monsoon closed in. Monsoonal weather is marked by very heavy rain and sweeping winds, which produce a complete white-out. The weather is usually too rough to go over, and we didn't carry oxygen anyway.

We were flying at about two to three thousand feet when we got into this storm, and we couldn't see a thing. We couldn't go up into the cloud because it was too dirty and too rough. Consequently, we ended up amongst the hills of the Ghats. I had the map out and was reading the topographical features. It was very very hairy.

He goes on to give a picture of navigation difficulties:
On our first trip we took off about 6 p.m. one night to go to Koggal. We stooged along all night and when we came near our ETA with about a quarter of an hour to go, we found we couldn't wake the CO who had come

with us. He had said 'Wake me up ten minutes before ETA. I have to show you how to go in through a special corridor.' The skipper and I were peering out; we could see a lighthouse on a little rock but not the landfall for our corridor.

Then, right on our ETA, with no help from the CO, we saw our pinpoint.

We then turned up the east coast. We used dead reckoning and astro. The D/F loops were useless—quite archaic. We used Loran [long range navigation] but it was in its early stages and not a lot of use.

We went up the coast to Calcutta and went into the Hoogli.

Just as we were ready to get out of the hatch after landing, the CO stuck his head down and said, 'Navigator! How far is it to Akyab?' I said to myself, 'That's in Burma.' I got out my maps and charts and said, '186 miles sir.' 'Good! We take off at 8 a.m. tomorrow.'

We had thought we were in for a break from flying but went off the next morning and put down at Akyab. It was a Japanese-occupied area, an island separated from the mainland by a river. We landed on the river and moored. The CO, the skipper and the squadron leader took off in a dinghy and left us on the plane. We dived in and had a swim. We didn't know what was on land—it was supposedly Japanese occupied.

He continues:

Living conditions were rough and crude. We lived in tents most of the time. We had a bearer, a boy who came and made the beds each morning and shaved you if you so desired. The tents had dirt floors.

I remember one stay in Calcutta. We were put into a big building overnight. We used to carry our bedrolls, wrapped up, in the plane with us. We didn't know much about the premises but in the morning we were scratching furiously. It had been riddled with bedbugs.

Back in Karachi, I had the bedroll out in the sun for days on end before they were eliminated.

The food was mainly curry and rice and, in general, wasn't too bad. It was fairly basic. No grog, gin sometimes, but beer virtually never. I remember a keg of Bangalore Bitter arrived one night. I took a mouthful and spat it out. It tasted just like vinegar. Some of the old diehards sat around drinking it just because it was called 'beer.'

David Nesbitt converted to Thunderbolts in southern India in preparation for operations in Burma. He had this experience:
The Thunderbolt had a greater range than the Hurricane and could penetrate far deeper. It had a greater ceiling and could fly over a lot of dirty weather and could carry a heavier bomb load. It could carry three 500-pound bombs or two bombs and a 90-gallon belly tank or one bomb under the belly and a 90-gallon tank under each wing.

About sixteen Hurricane squadrons converted to Thunderbolts. They were provided to the RAF under the wartime Lend/Lease program. I did my conversion in southern India and had an experience I will never forget.

I had about thirteen hours on the new type and was practising aerobatics. I had done one hour in the morning and took off for a second hour in the afternoon. I climbed to 32 000 feet and did a loop, climbed back to 32 000 feet and started a roll off the top. I must have pulled it too tight because the aircraft started to mush and then did a vicious flick and went into an inverted spin.

Inverted spins are dangerous and were not taught in the RAF or RAAF flying courses. I recognised this one because my instructor on Tiger Moths in Australia had given me an unofficial demonstration.

Doug Nicol's barracks, Korangi Creek, Karachi.

Doug Nichol's tented quarters, Red Hills, Madras.

The centrifugal force throws you towards the top of the cockpit and I found I could not get my feet on the rudder pedals and couldn't reach the latch which would jettison the hood and bale out. I was trapped.

However, the aircraft eventually stopped spinning of its own accord and I was in a screaming dive with the aircraft literally standing on its nose and slightly inverted. This was controllable so I eased the aircraft out of the dive. I noticed two things. The instrument panel was heavily coated in oil and I was at 2000 feet having lost 30 000 feet in the spin and dive.

When I landed I found the cockpit was covered in gushing oil and I had gashed my elbow in the initial flick. The Engineering Officer eventually found that the oil tank, which was bolted down forward of the instrument panel and the firewall, had broken loose and had been pierced by a protruding rod.

Shortly after another pilot, from another squadron, got into the same situation. But he was flying a Mark II Thunderbolt (Mine was a MarkI) and the emergency hood release was in a different position and he was able to bale out.

Aerobatics was part of the program to familiarise pilots with the feel of the aircraft in different situations so that, in an emergency, they could quickly recognise the position and take corrective action.

BURMA

Operations from Indian Bases

• Calcutta

Ron Gardner describes the activities of 681 Squadron, on which he served:

I did all my flying in the east with 681 Squadron. When I arrived in Calcutta, 681 Squadron was being formed as a PRU flying Spitfire Mark XIs. It was to be based near Calcutta. It had forward air bases in the northern part of India, and, later, in the Imphal Valley and south in Burma, towards Monywa.

We had two pilots go down in the Chinn Hills, both with motor failure. One was Jack Strong from Melbourne and the other was a little New Zealand bloke, 'Pappy' (F.O. Papps) from Rotorua. They both parachuted down. Jack Strong walked back through the jungle. He told us that he was walking along a path when he came face to face with a big monkey. Neither uttered a sound and they both kept on walking.

Jack made his way to a native village where he insisted on eating the same food as they did. Poor old Jack was in hospital with dysentery for five days after he came

back. They carried him straight out of the jungle to the hospital. Fortunately for him it wasn't far from the village to the hospital.

F/O Papps got back under his own steam.

I carried with me in the aircraft as defensive weapons, a big Indian kukri, a sheath knife and a Smith & Wesson revolver. As we were at high altitude most of the time we wore the heavy flying suit which was normal in the UK.

For 60 per cent of the time the weather in Burma was very good for flying. But the monsoon weather was atrocious. We had one case where six Spitfires coming down from Calcutta to Puri for maintenance were lost in a monsoon storm. Two pilots were killed in crashes, one baled out but broke his back and the other three baled out unhurt and were picked up. The chap who broke his back couldn't get out cleanly and hit the aircraft badly as he left.

They had flown through a front and run into tremendous turbulence.

• Hathazari

Hathazari airfield was situated near Chittagong in India. It was occupied for a time by 31 Squadron RAF as a base for carrying out transport sorties into Burma. John Eather writes of an experience there:

We could be short of petrol on the way back. I was on my way out from Hathazari, just over the border. I was carrying thirteen or fourteen 44-gallon drums of petrol to be dropped by parachute. About three-quarters of the way to the dropping zone a motor started to pack up so I decided to feather it.

I then started to make a very gentle turn but couldn't hold height with the load on. So I decided to drop the petrol. By this time I was back in our own territory so I dropped a good half of it by parachute and plotted my position. They were always crying out about being short of petrol so I thought they could try to recover what I dropped. But I later learnt that they didn't try.

• Kumbhirgram

Bill Hughes was stationed at this Indian drome, which had a concrete all-weather strip, with 47 Squadron RAF at the beginning of 1945. From there he operated over Burma in support of the XIVth Army.

Bill tells of night-intruder sorties:

We were operating across the Chinn Hills down into northern Burma from Assam. We were initially using Beaufighters which had a fairly long range. We could fly for four to five hours which was useful for reconnaissance of roads, railways and rivers.

We did night-intruder work as well as daytime attacks

on areas like Mandalay and Anisikan where the Japs were re-equipping and bringing up supplies.

Getting down into Burma at night was quite a feat. I was lucky because I had done a lot of night flying while I was a flying instructor and this helped enormously.

Night flying in Burma was particularly hazardous as there was a build-up of cu-nim clouds in the afternoons so that it was often almost impossible to get back because of the beating you got in the clouds over the Chinn Hills. There were times we had to land at advanced strips like Magwe.

The cu-nim had violent downdraughts in them. They could take you down 3000 feet at the drop of a hat and you could do nothing about it. Having taken you down it would bottom out and then take you straight back up again. It was a hell of a shock. This meant you had to be at 12 000 feet to cross the Chinn Hills in case you got a downdraught. There was no way you could go over the top of the cu-nims. I have had a Mosquito at 20 000 feet and still in the cloud. Some of them went up to 60 000 feet.

Nor could you go round them. On these long-range trips you often found yourself short on fuel and you couldn't afford the extra distance.

At night you could not see anything but the lightning. There would be a tremendous noise from hail striking the aircraft—it was really frightening.

I did one trip on Mosquitoes and I was quite certain my hair stood on end. I had no control at all over the aircraft. The instruments were bounding hither and thither and I don't know how I stayed level or how I got out of it.

I got back one night after being beaten up with hail in a cu-nim to find there was no paint left on the main plane or the propeller bosses and the aircraft was pitted. The Engineering Officer wanted to know what I'd done. He took some convincing it was the weather.

We lost people through the weather. I was glad I had the experience of a thousand hours as an instructor. I would not have wanted to have come straight to it from an OTU.

While it was generally easier to find your way round in Burma than in England with all its smog and smoke it was a great help to have a good navigator to get you close enough to a target that you could pinpoint it. At night the only thing we could do was try to spot the lights of trucks and trains.

I often wondered how we got back from some of our trips. We followed a set course and didn't have much radio help. There were the lights of the aerodrome at the end but the rest was pitch black.

Retreat into India

The Japanese Army crossed the south Thai border into Burma on 20 January 1942. The British troops in Burma consisted of the 17th Indian Division and the 1st Burma Division (each short of one of their three regiments). In all there were about 2000 men.

The 17th Division was given the impossible task of stopping the 15th Japanese Army and the Burmese division was left in the Shan states in the north of Burma.

The air defence was pitifully weak. Ten RAF Buffaloes and fourteen American Volunteer Group Tomahawks had to take on greatly superior Japanese forces.

Rangoon fell to the Japanese on 9 March 1942 and Slim was appointed to command the British corps in Burma. He immediately tried to unite his forces and staged a fighting retreat along his line of communication to India. He was hindered in attempts to mount a counter attack by the refusal of Chinese forces to advance beyond Toungoo.

On 21 March almost all of the available aircraft were destroyed by a Japanese raid at Magwe. The remainder were flown out to Akyab and were withdrawn to India shortly after.

With a grave shortage of aircraft the RAF attempted to support the retreating troops from remote airfields immune from enemy attack.

Spiro Tsicalas tells of his involvement:

After we arrived in the Middle East we went into an RAF Aircrew Pool at a base on the canal. We were sitting around there for about three months. We were then posted down to Kenya to a Blenheim OTU at Nakaru.

We crewed up and did the course. We used to fly on cross country flights around Mount Kenya and Mount Kiliminjaro and up as far as Abyssinia. I crewed up with an RAF sergeant pilot and a South Australian navigator.

After doing the course we went back to the Middle East to the Aircraft Pool. We were there a week and then we were sent to Wadi Natrun to pick up a Blenheim and fly it out to the Far East.

We got a very clapped-out Blenheim on which nothing worked. Both engines were dropping at least a couple of hundred revs, the turret wouldn't work—it used to creep, the radio was u/s and the bomb sight was broken.

We went out via Haifa, Iraq, Kuwait, Bahrein Island to Karachi. We complained about the aeroplane to the maintenance people there. They said they'd have a look at it if we'd give it to them for a few days. They put it in a hangar and did all sorts of things to it. When we tested it a few days later it was even worse.

We were told to take it on to Allahabad. We were met there by a Wing Commander Cox. He said, 'Where are you going with this aircraft?' We said, 'Rangoon.'

The heavily armed B-25 Mitchell bomber.

He replied that Rangoon had fallen and 'I'm keeping you here to join my squadron No 34.' He had picked up three or four other crews and now, 'I want to get operational.' The squadron had been in Singapore for about two years before the war but had been forced out by the Japanese. They lost all their aircraft at Tengah [Singapore] and Rangoon [Mingaladon] on the ground. He got out by hitching out accompanied by some crews.

We joined 34 Squadron about April/May 1942 and started bomber operations almost immediately.

Our first mission out of Allahabad was to Burma— to a place called Pakokku which was situated in a valley, as was Imphal. There were about eight Blenheims on this mission. But the valley was completely covered in cloud so we aborted the mission and went back to base.

It was decided to try again the next morning but the same thing happened.

A squadron of B-25 Mitchells had arrived from the United States about a week before this. We hadn't seen one before and they looked magnificent alongside our old clapped-out Blenheims.

It was decided that they should carry out the mission with their sophisticated bombsights and other equipment.

We gathered on the airfield and watched them take off. We expected they'd be back in about four hours. Away they went and we waited and waited but they didn't return.

Apparently they got over the target, went down through the cloud and didn't come up. As far as we know all eight aircraft were lost on this their first mission.

A fortnight later 34 Squadron moved to Ondal in Bengal which was a new base. We were the first there and we carried out day and night operations into Burma. Our targets were marshalling yards and airfields at Akyab, Mandalay, Magwe and Shwebo. Quite a few shipping strikes into Akyab Harbour were also included.

The Japanese ships would come into Akyab harbour and unload supplies there. Our crews would go in the Blenheims at low level across the Bay of Bengal—which I assure you is not to be recommended.

The wharf area was very heavily defended with small arms fire. As well, there was a Japanese airfield in the vicinity with Zero fighters on it. It wasn't mentioned in our briefings. The intense ground fire led quite a number of our aircraft to fly straight into the wharf and blow up.

There were two other RAF Blenheim squadrons (No 113 and No 60) near Asansol. There were several

Australians on both of these squadrons. Quite often the three squadrons would combine on a big operation of eighteen aircraft to prime targets such as Mandalay.

We carried on doing these operations until April/May 1943.

The diciest trips were the shipping strikes. We went in at low level all the way and there were generally many Zeros around the area when we arrived and the ground fire was intense around the docks. We would be at about forty feet above the water. Zeros made passes at us as we left the target and we kept low across the Bay of Bengal to evade them.

When we first operated there were virtually no ground radio facilities. There was a medium frequency station at Calcutta which was very underpowered. Other radio facilities may have existed but were unsuitable for our operations. Also, the equipment in the Blenheims was antiquated and was often unserviceable due to lack of spares.

We had a prang early in 1943. We burst a tyre on take-off but went on with the mission. The plane skidded badly on landing and burst into flames. It was a write-off.

Bit by bit retreat became the order of the day. The rich oil fields at Yenangyaung were destroyed on 15 April—a bitter blow.

Nine hundred miles of fighting retreat brought the Burma Corps to the Indian border. At Imphal the weary 17th Division and some elements of the Burma Division prepared to resist further Japanese advance.

• Arakan

The Arakan is in the north-west of Burma and borders the Bay of Bengal. The Mayu Range is close to and parallels the coast providing a very narrow coastal plain rising to very steep jungle-covered hills which are difficult to traverse and a coastline cut by innumerable streams.

There were three campaigns to recapture the Arakan but I will only deal with the period from late 1943 to 1945.

The main aim was to recapture Akyab to provide a port and airfields to assist in operations in central Burma as well as keep some Japanese forces from intervening in those operations.

The second campaign began on 30 November 1943 with the advance of the British XV corps and provoked a strong Japanese reaction which resulted in the British being surrounded in two defensive boxes cut off from all supply except by air. A violent battle was fought for two or three weeks until the Japanese fell back exhausted.

Spiro Tsicalis, now flying in Dakotas, had a lucky escape during this battle:

General Slim's XIVth Army was many times on one side of the dropping zone and the Japanese on the other.

We dropped in the middle, no-man's-land, and sometimes the Japanese got part of the load.

On one occasion we were dropping with three other squadrons (Nos 62, 194 and 117) with ten aircraft in a big circle. When we were running in for our final drop the pilot rang the alarm bell. I raced up the front and he said, 'Look at those bloody Zeros.' There were twenty or thirty Zeros in amongst the transports so close you could see the red and green paint. We did have an escort of Spitfires and Hurricanes but they were up at about 20 000 feet.

The pilot said, 'Get the rest of the load out and we're off.'

I was standing in the door getting the last of the load out when an American Curtiss Commando went right across us. He was so close I could have grabbed his tailwheel—I think he was on his first trip. He dropped about twelve chutes [parachute loads] of canned milk. Our pilot was frantically dodging the chutes but we collected one in between the starboard engine and the cabin and many cans of milk came through the fuselage like cannon shells. One chute tore into the leading edge of the starboard wing.

The pilot told us to put on our chutes and we skimmed over the jungle with one Zero chasing us. He gave up after a few minutes.

Five or six RAF Dakotas and three American Commandos were lost or seriously damaged in this action. When we got back to Agartala all the ground crew were amazed at the great holes in the aircraft.

While we were standing on the airstrip an American Dakota returned from the mission with no fin. When he landed I spoke to the young lieutenant. He said, 'Those goddam Zeros attacked me. I turned into him and he made a head on dive at me. I put the nose down and he slid across the top of my plane and took off my fin. You can see the red and green paint from the Zero on my aircraft.' The paint was there.

In December 1944 the British XVth Corps renewed attacks in this area. On 2 January 1945 it entered Akyab. The capture of Kangaw followed, as also the islands of Ramree and Cheduba. Thus sea-supplied airfields became available.

David Nesbitt flew Thunderbolts in this area in December 1944:

We were sent to the Arakan which is on the coast of Burma. When we arrived at Nazir we found a grass strip bulldozed out of paddy fields and jungle about thirty miles south of Cox's Bazar. Cox's Bazar was an all-weather strip with bitumen runway and taxi strips. It was used by us and other Thunderbolts when the grass strips were unserviceable.

Our first duties on arrival were escorting DC-3s on supply drops in support of the XIVth Army in the Arakan. This was on the coast as distinct from the central front. Another type of operation at this time was escorting RAF Liberators and US B-24's on long-range missions to targets at Rangoon and beyond in the Malay Peninsula. These usually involved a round trip of four and a half to five hours.

The cockpit of a Thunderbolt was very spacious but the pilot was secured to the seat by his Sutton harness and his whole body movement pivoted on his coccyx (tailbone). We would be lying on our stomachs for the rest of the day because we couldn't stand any pressure on our tailbones.

The escort operations in which I was personally involved were medium cover. There were three types of cover; close, medium and top.

Close escort flew within sight of the bombers. Medium escort was three to five thousand feet above the bombers and top cover was well up out of sight—it was generally provided by P-38 Lightnings. The top cover generally did a sweep before the main force.

After the target had been bombed there were generally another couple of squadrons to provide cover during the withdrawal (four and a half to five hours).

These operations were to targets in and around Rangoon and to railyards at a place called Mokopolin which was a big rail junction. Another target was Nopthalon, a rail target which was at the end of the infamous Burma Railway. There was considerable anti-aircraft as well as Jap fighters located there.

Kohima

At the beginning of April 1944 the Japanese cut the Imphal-Dimapur road at Kohima in their thrust to India. In accord with Slim's dictate of October 1943, the troops stood firm where they were and were encircled.

The besieged garrison numbered approximately 3000 men. They were besieged by more than 6000 men of the Japanese 31st Division.

The garrison was supplied entirely by air for the twelve days until they were relieved and the Japanese were driven off.

Spiro Tsicalas tells of his experience:
One of the most important battles took place at Kohima in early April 1944. Kohima was a little town located in the Naga Hills about sixty miles due north of Imphal and was a key point on the only road leading down into the Assam Valley.

These troops held out against all odds. Supply was entirely by air. [The dropping zone at one stage was the size of two or three tennis courts.]

We dropped there three or four times a day to provide food, water, mortars, ammunition and petrol.

I understand that 31 Squadron was the only squadron used to supply Kohima because we were the most experienced. There were about eighteen aircraft in the squadron at the time. We flew from Imphal and our main base at Agartala.

I can recall standing in the door on one occasion after about our third drop. I could clearly see bodies in the street. Suddenly there was a rattle of machine-gun fire. We were riddled. One bullet hit the jungle knife I wore in my belt and damaged it.

We got in about 60 per cent, sometimes 70 per cent, of the load despite the smallness of the drop zone.

It was dangerous for the ground troops to retrieve the loads. On one occasion a soldier raced out to try to catch a one hundred pound bag of rice and was badly injured.

Kohima was finally relieved on 20 April.

Imphal

Imphal was only seventy miles from the border between Burma and India. It is situated in a depression in the mountains, forty miles by twenty miles in extent, called the Imphal Plain. The surrounding peaks reach from five to ten thousand feet above the level of the plain.

It was established as a supply and administrative base widely dispersed against air attack and not planned for defence against ground attack. It was connected to the railhead at Dimapur by a good (and only) road via Kohima.

The Japanese attacked in March/April 1944 with three divisions as the start of their thrust into India. But they had little armour and had lost control of the air through the Spitfire's superiority over the Zero. A crucial factor was the complete inadequacy of their supply arrangements which forced them to rely on captured supplies for their sustenance.

Ron Gardner describes the activities of 681 Squadron on which he served:
I was posted to the Imphal Valley. The airstrip was right on the fringe of the front line. The Japanese were on the hills just outside Imphal and used to rake the whole area with howitzers and other artillery fire. The infantry were just outside the boundary of the valley. That is as close as they got into India.

The strips were not bad. In the Imphal Valley the main airstrip was a simulated macadam surface. It was pretty good. We had to have a hard surface with our heavy load and small tyres.

In Burma I did a lot of low-flying reconnaissance with the Mark VC Spitfire (with the clipped wings)—intelligence flying. I was engaged on general work. It was heavy jungle country and I had to do a lot of feature-line overlaps and mosaics and pinpointing of the area.

In one case we had to photograph, in a ten mile strip, ten or eleven bridges over one river. The army had to know which bridges were intact.

I also flew high-altitude Spits. These were painted a special high altitude blue which made them invisible from the ground at 15 000 feet or higher. The whole leading edge of the wings, from main spar to leading edge, was a petrol tank which was filled through filler caps right up at the wing tips. We could fly for five hours and ten minutes in these aircraft; and cover a lot of miles. A lot of my trips were only an hour and ten minutes duration but every now and then there would be one of three, four or five hours.

The pilot had to do his own navigation—it was all dead reckoning. I had a folding map which I could put on my knee pad and open out like a road map, changing to different sections as required.

Burma was the easiest place I ever had to map-read over, there were a lot of easily distinguishable landmarks. Once you got into the country the rivers, the shape of the jungle, the big towns like Rangoon or Mandalay made it easy. One of the most prominent landmarks was Mount Poppa, the home of the king cobra snake, and not the place to lose the motor. But coming back across the Chinn Hills to Calcutta meant flying over 400 miles of jungle and any pilot who went down in that had a hard time.

David Nesbitt, an RAAF pilot with 5 Squadron RAF at Wangjing, continues the story:

I arrived in India in October 1943. We operated from Assam to start with. At the time Imphal was surrounded we had a forward detachment based on the Imphal Plain. We operated from three airstrips: Wangjing, Palel and Kangla. All had been bulldozed out of rice paddy fields and, with the exception of Palel, which had a bitumen runway, had earth and gravel runways.

The Hurricanes did very well on those runways. We had no maintenance problems in my time.

We were inside the Imphal perimeter which was manned by British, Indian and African (East and West) troops. Within that perimeter we retired at night into what were known as boxes. These were in the form of trenches, dugouts and sundry bashas surrounded by barbed wire. We were guarded at night by Gurkha troops. The planes were in what were called revetments and they also had a Gurkha patrol looking after them.

During the siege we were anything from seven to fifteen miles from the front line. When we were operating from Palel we were receiving something like twenty-two shells per night from a mountain gun which the Japs had manhandled to a high peak. I found out after the war that they dismantled and hid it during daylight—which is why we never found it by air reconnaissance.

We engaged in several types of activities. The primary one was what was known as a rhubarb. It was usually performed with two, sometimes three, aircraft. We flew into Burma, well behind the Japanese lines, and flew above the roads at about fifty feet. We weaved from side to side looking for Jap transport and/or troops concealed in the jungle. The Japs were past masters at the art of camouflage and concealment.

If we found any Jap transport, either concealed or, very rarely during the day, on the move it was attacked. We used four 20 mm cannon.

My number one on my second ever operation on Hurricanes was what was known as a British Latin American volunteer who had won the DFC. These volunteers were usually sons of Englishmen who had settled in the Argentine and were often connected with the big meat companies. Often they had been educated in England and volunteered for the RAF.

We were doing a rhubarb when we spotted a Jap staff car. My leader immediately called over the R/T. We had seen it simultaneously and started to climb. We did a circuit round the car and climbed to 1500 feet. My leader went in first. He dived and, as he started to pull up, he rolled over on his back and went in upside down. Yellow flames and smoke swirled up from the jungle as he hit the top of the trees—he exploded. I can still see it vividly fifty years later. The return fire wasn't visible.

I sensed that it may have been a trap and broke off. As it was only my second trip I didn't know where I was. I knew that if I flew due west I would hit the Chindwin River and this I did.

There were a lot of fires in the jungle and this produced the same effect that we used to get in Australia with bushfires. I stayed at about 150 feet in order to get pinpoints and map-read to get some idea where I was. Fortunately, when I hit the Chindwin River I was able to identify some small islands (really sandbars) which appeared on my map and establish where I was.

I circled the island while I climbed to 14 000 feet to clear the Chinn Hills and called up for a homing on the Mayday frequency. The controller gave me a vector to fly and it brought me right over the top of Wangjing, where we were based.

There was turbulence but not so much in the morning. This rhubarb was about two in the afternoon and in the heat and tropical conditions the aircraft was bouncing around all over the place.

Our other major activity was close support for the army. At that time we carried two 250-pound bombs, one under each wing. We were vectored on to the target by smoke bombs fired by the artillery.

We released our load from sixty to one hundred feet. A lot of fellows did get caught in the bomb blast. Some didn't pull out in time, or not properly, and wound up with a tree or something caught up in the radiator (it was under the belly of the aircraft in a Hurricane). They were either lost or just struggled back home because the branches or leaves blocked the air flow and the engine overheated.

These attacks were made in very rugged country where the Japs were in valleys and gullies or, often, between two peaks. We lost a considerable number of aircraft.

We lived in tents and bamboo bashas at Imphal and at one stage lived in trenches with sandbag dugouts—very much like the pictures I have seen of the Western Front in the First World War.

The food was terrible—mainly bully beef and dehydrated potatoes.

There was a high incidence of malaria although it was being brought under control through an anti-malarial program which involved taking Mepacrine. This suppressed the malaria if you were bitten by a malaria carrying mosquito. But if you stopped taking the tablets the malaria became active. Prior to this program the rate of people going down with malaria was high. They had to be hospitalised.

Imphal was completely surrounded and everything was in the firing line. The Dakotas and Wellingtons which flew in the supplies flew out the casualties to hospitals in India. They flew down the corridor.

My longest trip in a Hurricane was two hours and twenty minutes. That was 200 miles into Jap territory. As we were following roads, such as they were, which meandered all over the jungle, we covered rather more than 200 miles weaving up and down hills and from left to right. We usually flew twice per day.

Aircraft maintenance was very difficult. The ground crew had to carry it out in the open, there were no hangars.

When we were surrounded the ground crew were very close to the Japanese lines and were shelled like everyone else.

I was in Imphal for five months, after which we were withdrawn.

Operation Thursday

The specially trained Chindit force was developed for long-range penetration behind enemy lines. It would march or be dropped by glider into the Burmese country-side.

It was given the task of implementing Operation Thursday. This was the strategy propounded by the brilliant but erratic General Orde Wingate.

His overall brief was to impede the provision of supplies to the Japanese 18th Division, to assist the Chinese advance across the Salween River and to generally harass the Japanese in Northern Burma.

Wingate therefore set these three objectives: (1) The Bhamo-Myitkyina Road; (2) The Railway; and (3) Indaw.

He also developed a system of fortified bases (strongholds) which would be defended by a garrison and would provide a means of collecting casualties and reinforcements.

Two strongholds were to be developed, Aberdeen and Broadway in the area of Indaw.

Spiro Tsicalas had first hand experience of these:
All the airstrips were dirt strips and presented problems in bad weather. We used to land there with donkeys, cows, mules and other supplies. Quite often the British Army held one end of the strip and the Japanese the other when we landed. It was as close as that. We would often be holed by small-arms fire so we'd waste no time in getting off again. The army unloaded everything without delay.

We were very vulnerable as we came in to land and quite a number were shot down. The Dakota squadrons were Nos 31, 62, 117 and 194. They were supplemented by American squadrons flying Curtis Commandos (C-46) and Dakotas. The C-46 could carry a huge load and the Dakotas were often towing gliders as well. Many gliders crashed on landing because of the difficult terrain.

Spiro Tsicalas also supplied columns directly:
We would supply drop to them as they were totally air-supplied. Everything was dropped including ammunition, petrol in drums, mortars, etc.

Food was probably the most important item dropped. It included rice, canned fruit, canned meat M&V. If we failed to achieve a drop they didn't eat.

Free drops were only used for things like rice—which was in double bags to reduce spillage. All other items were dropped by parachute.

The army liaison officer on the squadron was in radio communication with the Wingate force and would give us a briefing. He would indicate the area of a drop on a map (e.g., the bend in a river). The army would put out a large white T or cross on the ground to indicate the actual drop zone. If there was no appropriate indication on the ground we would abort the drop.

The enemy was so close to the army position it would occasionally benefit from the drop. Generally the army received 60 to 70 per cent of the drop.

We went at staggered times—about a quarter of an

hour between aircraft. We took off with about 8000 pounds of material. The Americans who later joined us in supply dropping considered we were crazy as they only took 5000 pounds.

The aircraft were loaded by an Indian Army company without training in weight, balance or correct positioning of the load. Sometimes we would be taking off and our pilot would say it wouldn't take off because it was overloaded. We would stop at the end of the strip and dump some of the load.

The navigator, second pilot and I would be responsible for dropping the load at our destination. We went down the back and took off the door. One would push with his feet and the others would put their shoulders to the load and out it would go. A chute only caught on our tail once and it did no harm.

David Nesbitt was flying a Hurricane and saw things from a different perspective:

I flew over Broadway (one of the strongholds) but didn't land. The Dakotas went in but our job was to provide close support to them. As soon as they got to Broadway we broke off and flew back. The Dakotas would hopefully be covered by another flight of Hurricanes for the trip back. Some were lost as they landed.

A technique was evolved whereby an air corridor was developed which ran roughly from Imphal to Broadway. The Dakotas flew in that corridor and were escorted by Hurricanes and, hopefully, a top cover of Spits. In my experience two Dakotas would usually be escorted by four Hurricanes but sometimes there were only two. There were both American and RAF pilots on Dakotas.

The Japs had discovered where the strip was at Broadway and aircraft engaged in that operation deep within enemy territory were very vulnerable to Jap fighter attack.

David Corthorn flew on one supply mission to the Chindit forces:

We flew on a supply mission to the Chindits. A lot of our Flight had been doing these runs for some time. Ours was a quite uneventful trip.

We were one of the last aircraft to reach our destination. We dropped the containers from about 1500 feet. There were kitbags full of eleven-pound tins of bully beef. Sometimes the bags split. We only made the one drop—there was a red cross on the ground and the Chindits were supposed to run out after the drop and collect the goods. Some people were over-anxious and were hit with the drop.

Spiro Tsicalas has some comments on General Wingate:

I actually met General Wingate. In September 1943 I flew from Bangalore to Calcutta in a Lockheed Hudson aircraft. We took off from Bangalore on what was to be a five-hour trip to Calcutta.

The only passengers in the aircraft were General Wingate and myself. I was a flight sergeant at the time. General Wingate was wearing a battered old topi and a bush jacket. He wore the ribbon of the DSO and the Palestine Medal with two or three rosettes on the DSO.

He carried a small bag affair which he opened after a couple of hours. He hadn't spoken to me at that stage. He then glanced across at me and said, 'Sergeant. Would you like to share my sandwich?' He had the biggest sandwich imaginable made of a loaf of bread and a couple of pounds of cheese. When I answered, 'Yes.' he broke it in half and gave one half to me.

He was medium build with long hair and had what appeared to be a small alarm clock strapped to his wrist. He was considered at that time to be very eccentric.

Hurricane aircraft. (Hurricanes used in Burma featured a slightly different air intake.)

Advance to and Crossing of the Chindwin

On 6 August 1944 General Slim gave XXXIII Corps the task of capturing Kalemyo and establishing a bridgehead across the Chindwin River. This was reinforced when Mountbatten set General Giffard, CIC 11th Army Group, the task of securing a bridgehead across the Chindwin by mid-December. This directive was passed on to Slim.

The road along which supplies had to travel was long, winding and narrow. It had to be widened to be of much use for tanks and their transporters and other heavy equipment. To add to the difficulties it was the middle of the monsoon season with continual torrential rain.

Beyond Taukkyan supply had to be entirely by air and would involve landing at temporary airstrips or dropping supplies.

Progress was slowed by seas of mud—particularly in the Kabaw Valley, but by early November large-scale moves could be made into the valley and Kalemyo was captured.

John Eather of 31 Squadron RAF (Dakotas) was involved in these operations:

We were suddenly recalled to Argatala in Bengal. We were flying C-47's—the military version of the Dakota. We arrived on 1 November 1944 with the rest of the squadron.

We then started supply-dropping operations in the Kabaw Valley, just south of Imphal where there had been a massive battle. On my first three trips I was screened by a very experienced pilot who sat beside me and gave me a quick rundown on what to do. The weather was pretty good so the flying part was easy.

We carried loads up to 7000 pounds varying from ammunition to toilet paper as well as petrol in 44-gallon drums—thirteen or fourteen drums at a time. Sometimes we'd drop by parachute (double chutes were used) and sometimes by free drop. Rice, for example, was free-dropped. It was loaded in double bags—one bag inside another—so that, if the inner bag burst, the contents were not spilt everywhere.

The aircraft flew in a circuit round the drop point. There could be anything from ten to thirty aircraft in the circuit. We dropped parachutes from about 200 feet. If we were in the mountains we dropped the plane into the valley, dropped some parachutes, climbed up, went round and repeated the process.

There were no strips in the Kabaw Valley so everything was dropped. We didn't get strips to land on until we got over to the plains. We were dropping to troops in local areas, not into a perimeter like Imphal.

Sometimes the loads fell on people, particularly the coloured troops. I've heard they tried to catch bags of rice and were flattened by them.

We rarely flew in formation. Once we took off we were on our own until we joined the circuit.

Jap fighters caught us over the Kabaw Valley on 8 November. There was no fighter escort that day. I heard a Mayday call just as we finished our drop and was able to get out. One of our crews was shot down. The pilot survived but two crew members were killed. It was the first of the fighter attacks.

The weather got worse and worse but the squadron took off every day provided we had visibility of 100 to 150 feet. Mount Victoria was 12 500 feet and you had to get up to 14-15 000 feet to get over it. The highest I got to was 16 800 feet—but only for twenty minutes as we didn't have oxygen.

The Chindwin was crossed in December 1944.

Advance to the Irrawaddy

In December 1944 the Japanese 15th Army and the 53rd Division retreated towards Thabeikkyin. The enemy also moved troops from Monywa to Pakokku.

The British pursuit of these forces was hampered by the trackless nature of the country and the hindrance to transport of supplies from the heavy rain.

The last phase of the approach to the Irrawaddy began in mid-January, with one column thrusting towards Meiktila in complete secrecy by strict radio silence and complete sheltering from aerial observation.

Having overcome its transport difficulties, XXXIII Corps also advanced protected by 221 Group RAF who gave its fighter support exclusively to it.

Air supply was given by the Dakotas of 177 Transport Wing.

Dakota—military version of the DC-3.

John Eather tells of the difficulties of operating Dakotas in this period:

When we got to more level territory the troops would bulldoze out a strip and we could land. This let us carry a full load and avoid the losses associated with dropping. Some dropped supplies simply were not recovered and the extra packing required for dropping made them bulkier and heavier, reducing the actual quantities carried. Our maximum load was 7000 pounds.

Some of the strips were just big enough to land or might have massive trees at both ends. If you misjudged you had to do a full throttle landing and then cut the throttles and drop.

Some strips were laid with mesh later but dirt strips had a dust hazard. It was all right if you were first in but after that the dust didn't quite clear between landings.

We took off and landed on the same runway. We'd unload and take off in the reverse direction. Some strips had an army field controller.

Besides the dust there were other hazards. The surrounding terrain could be heavily timbered or, if the strip was a regular one it could well have been bombed and the craters were not always properly filled.

An unusual hazard was provided by shite hawks. They had a wing span of up to sixteen feet and we met them over areas where battles had been fought. They were where the bodies were. They were big and if they hit a motor when you were flying low—we often flew at two hundred feet—you were in trouble.

The rest of the crew pushed out the loads but I liked to have the co-pilot with me in case I got hit.

Sometimes we carried army fellows—they were always keen to have a flight. They would stand there with no straps on—quite eerie.

We didn't get much leave. I had leave twice in seven months. We'd send an aircraft every weekend for supplies and two or three crews would usually go down for leave. The main supplies brought back was Carew's booze for aircrews. We had plenty of that.

Some of John Eather's trips involved the carrying of passengers. He explains:

I carried a lot of troops one way and another, English, Scots and Gurkhas as well as East and West Africans.

I carried a lot of African troops. They'd try to put their hands out of the windows to grab the clouds. But they were very well disciplined. For example, I would simply tell the officer in charge what I required as regards smoking and there was no way those instructions would be disobeyed.

We carried wounded who couldn't walk out. They were on stretchers lashed to the floor—we couldn't take more than two or three at a time. There was usually a medical orderly with them.

I once brought out some Japanese prisoners. They had a couple of guards and were no trouble. However, on the same day, another fellow in the squadron brought out three Japs and their guards were violently airsick. They were lying there with their Tommy guns discarded on the floor. The pilot sent the wireless operator down as soon as he found out but the Japs didn't bat an eyelid.

Because of this incident I refused to carry prisoners the next time I was asked and told the sergeant major why. It went right up to a colonel who told me to carry out my orders. I said, 'You have no power to order me.' I reported this to the Commanding Officer. He said, 'Forget it.'

The coloured troops were good at unloading but had to be watched. They were likely to roll a 44-gallon drum of petrol out and let it drop six feet to the tarmac. After all, we did have special ramps.

I used to get off again about ten minutes after unloading. There were always little tents set up where you could get bloody awful tea and a bit of weevilly fruit cake. Many times I couldn't get back to base for a midday or evening meal and I used my K rations. It helped if you could get something hot. I could take off at 4 a.m. and not get back till 8 p.m.. They were long days.

Bill Hughes, who flew Beaufighters in the latter stages of this campaign, had an experience flying Gurkhas in a fighter bomber rather than in a transport:

The Gurkhas were very active in Burma. The Japanese hated the sight of them because they would charge right in. They were very tough characters.

When we were at Kumbhirgram the 7th Gurkhas were pulled out because they had had a very torrid time. They had been beaten up by aircraft from both sides while in close encounter with the enemy as it is hard to know who is who in the very close-range fighting. So they hated the sight of all aircraft and used to fire at anything that came over.

Their officers were very concerned about this and came down to the squadron. They thought it might be a good idea if some of the Gurkhas got to know some of the air force people. They suggested that some of the NCOs be shown over the base, see some aircraft and be taken on a sortie.

The CO said to me, 'Would you take one of these NCOs up on your next sector recce?' I said, 'Yes.' and this fellow turned up. You never saw such a thickset fellow. He was tough with a big kukri (Gurkha knife) in his belt. I thought, 'This bloke's going to be behind me holding on to that'.

We flew out and he was absolutely fascinated. Then we found a few trucks on the road. I went into the attack and he began screaming at the top of his voice. I thought, 'My God, I've got a real problem here. If he touches that kukri I'm going to lose my head.' But he was really wildly excited and very personally involved in the attack. When we got back I saw him talking to his men, explaining the dive with his hands and showing by action what he had experienced with much excited verbal accompaniment.

The upshot was that we were invited to the Gurkha mess for drinks—and that was fatal. It was pouring with rain and we were in a big marquee with Gurkhas drinking rum and water—mostly rum. They had what looked like pint glasses and the CO said to me, 'We're only going to be here for an hour—no more. We're going to cut it off, otherwise you blokes will be under the table in no time'.

While the rain poured down these grinning Gurkhas were bringing in great demijohns of pure rum. You had to keep face. There were many toasts—to all kinds of things. Every time you emptied a glass they'd fill it up again.

I managed, while they weren't looking, to pour it out on the sand floor but that didn't do any good because they simply filled it up again.

When we were all just about stunned the CO rose to his feet and said, 'Well, thank you very much, but we must go.' and we staggered back to our own mess very much the worse for wear but on very good terms with the Gurkhas.

Crossing of the Irrawaddy and Capture of Meiktila

IV Corps commenced the crossing on 13 February 1945. The RAF gave support in supply operations as well as close army support by Beaufighters, Mosquitoes and Thunderbolts which were based at Kumbhirgram and Wangjing. This involved long flights to their targets.

Bill Hughes gives an indication of what was involved:
We had Christmas at Ranchi in 1944 and then moved to Kumbhirgram. We provided low-level attack support to the 14th Army. We attacked specific targets—very few large targets but mainly roads, rail (shooting up trains) and vehicles on the roads—including bullock carts when we found a string of them. We found the Japs were using them.

There had been a ban on shooting up bullock carts because the Burmese were using them for transport. However I made an illegal attack on some and there were explosions among them. I thought I was gone because I had done the wrong thing but we got an order from Group the next day saying we could attack them.

We flew as low as possible, often at 100 feet, when doing recce work. Down the Irrawaddy you had to stay low if you wanted to find the Jap barges because they pulled into the sides where they were under cover. But you couldn't fly down the centre of the river because the Japs had tripwires across the rivers. We lost a few aircraft on these low-flying sorties.

Barges were hard to find unless you were given a specific target. The trees seemed to grow out over the Irrawaddy and there was ample opportunity for the barges to pull into the shore when they heard the sound of an aircraft.

You were lucky to spot them, and if you did you were not in a position to attack. You had to make a turn and find the spot again, which was hard in the jungle with no pinpoints.

John Eather gives his experiences at Meiktila:
At that stage of the war Meiktila was the main Japanese fighter drome but we kept taking it. It was on the side of a lake and you never knew whether we had it or the Japs had it.

We usually had parachute stuff but were instructed that if the Japs had been driven off and we had the strip we should land because it was more efficient.

We'd take off about dawn. The British troops usually managed to get the Japanese off during the night but the Japs would get it back during the day. Sometimes it went the other way round. We'd get there just after dawn but we wouldn't know whether we were going to land or drop by parachute.

Usually the army had the Japs pushed well back but sometimes they would be filling in holes as you landed or there would be mortar fire.

It was quite a joke on the squadron—'Who's got it to-day?'

It was at this time that he had his hairiest trip, as he explains:
We were briefed that it was a fairly new strip. It was quite reasonable by our standards and there didn't seem to be any problem except that there was a lot of cloud. We were at about 12 000 feet when we got a radio message that the Japanese had been attacking the strip. All lights had to be extinguished and we would be talked down.

There were no lights on the strip but we had a jeep at each end of the runway and they flashed their lights. There was an American to talk us down.

We had a lot of aircraft in the circuit that night and we were stepped down from 12 000 feet by our landing numbers. Round and round we went. It was cold and I reckon I lost at least half a stone in weight handling the aircraft. It was a very eerie experience—it was pitch black. Once we got down low we could see the planes going in and they seemed all right.

They gave us an altimeter setting so we could adjust our altimeters but I couldn't understand the American's accent. He'd say, 'Up a little, down a little.' When, by guesswork, you reached the end of the runway you cut your motors and hoped for the best.

I couldn't see a thing as I came in. I just had to rely on someone else dropping me in. I hoped the second pilot, who was reading the altimeter, had made the correct adjustment.

I kept the nose up and kept plenty of power on. I'd lined up with the first jeep and then the second but their lights weren't on all the time and the American was saying, 'A little bit port' or 'A little bit starboard.' Then I dropped it in because I didn't want it to run too far.

It was the worst experience I ever had. It was worse than being shot at. I only did it once.

In 'Defeat into Victory' General Slim has this to say about the support given by rocket firing Hurricanes: 'proved our most successful anti-tank weapon and their best day was 20 February when they knocked out thirteen medium tanks' (p. 421).

David Nesbitt has an interesting supplementary comment:

The Japs had tanks but I personally only attacked a tank once. It came out in the open.

The RAF had a special anti-tank squadron which had been brought out from the North African desert. They were equipped with two 40 mm anti-tank cannon. A friend of mine was a pilot on this squadron (No 20) and he reckoned the aircraft nearly flew backwards from the recoil.

On 5 March IV Corps captured Meiktila and prepared for any counter-attack.

Attacks on and in the vicinity of Mandalay

By early March 1945 the 19th Division of XXXIII Corps was ready to drive south on Mandalay. 2nd and 20th Divisions had crossed the river to the west of Mandalay and were ready to attack, with the RAF ready to lend very active support.

David Nesbitt, flying Thunderbolts, was one of those who attacked:

Fort Dufferin in the city of Mandalay was holding up the 14th Army. We attacked it with two 500-pound bombs per aircraft and then strafed it with our eight .5 inch machine-guns (or 50 calibre as the Yanks call them). We did this frequently and were unescorted on these operations.

We operated daily when we weren't on supply escort. Fort Dufferin was an old type of building built like a medieval fort with thick stone walls. There were deep archways—semi-tunnels—in the walls and these gave the Japs good cover.

We dive bombed and pulled out at about 1500 feet.

Bill Hughes attacked Mandalay on 26 January 1945 in a Beaufighter:

I did an attack on Mandalay after my fourth operation. Four Beaufighters took off, each with eight rockets. The objective was to attack a large building on the bank of the Irrawaddy. We went in at first light—about 3 a.m.

I had an aircraft which had been built at a car factory and it hadn't been put together very well. It was about ten knots slower than the other Beaufighters. We were flying over the Chinn Hills at about 9000 feet when I lost track of the others in the darkness.

We had a little identification light so we could keep in formation but I couldn't keep pace with them and lagged behind. When my navigator asked me what I intended to do, I pulled the power back and decided to go on at my own pace. 'You never know, we might catch them,' I said.

They slowed down to get the timing of the attack right and I picked them up about fifty miles from Mandalay. Then they opened up again. The CO was leading and I couldn't keep up with him.

I watched the others go in. All of their rockets hit but I was still crossing the river expecting to be shot out of the sky.

For a rocket attack we went up to 6000 feet and then came down in a shallow turn, released the rockets and did a split-arse turn to avoid the debris. I sent all my rockets in and the roof and girders were flying round the aircraft. I could only see the splashes in the water from falling debris. I didn't see any flak—I was too busy. We didn't suffer any damage at all to the aircraft.

The building was estimated to house a thousand Japs as well as being used as a store. The idea was to hit them before they got out of bed. We never heard the result but the building was a wreck.

That was one of my few big targets.

John Eather recalls a trip over the Mandalay area:
Prior to the British taking Mandalay we had occasion to do a couple of nickel raids over there. We weren't too happy about dropping propaganda leaflets. They were to be dropped from low level at night—a mission which didn't seem very important to me.

We went in at a fairly high level. It was a nice moonlight night and we dropped them, taking account of the wind. We weren't all that fussed about where they went. The targets that night were Mandalay and Myingyan and Meiktila, the Japanese fighter base, was in the middle so we weren't too keen.

A few days after Mandalay fell we had occasion to take some civilians back to our base. They had been prisoners in Mandalay. One was a young Irish priest. I invited him up to the cockpit and we got talking. I asked him about that night and, believe it or not, the pamphlets dropped right on the target.

We flew into Mandalay the day after its capture. The army boys had fixed up the drome overnight—they wanted to eat. I was the first bloke to land in Mandalay after we took it.

After that the army started to move pretty fast. We carried lots of small-arms ammunition but not shells. The ammunition was in boxes. We used to carry stacks of toilet paper. We took in food too. It was put in hessian bags and pushed out.

The volume of work hotted up. Our tour was 700 hours and I did that in seven months. In the earlier days they only did fifty hours a month.

Bill Hughes was stationed in the area near Myingyan and operated from there. He comments on Kinmagon from an operational point of view and as a place in which to be stationed:
We moved down to the plains in April 1945. That brought us closer to the scene of activity and we could operate round places like Rangoon. Prome was always a very hot spot.

It should be remembered that the Japanese put over 200 000 men into Burma. People think it was a small campaign. The Americans later indicated that they were very concerned about Burma because, if the Japs had broken through into India, there were a lot of pro-Japanese Indians and the oilfields could have been vulnerable. That's what the Japs were after and it was touch and go for a while.

Kinmagon was just a strip bulldozed out of the plains but it was an all-weather strip. We lived in tents and had to bring our own equipment down in a limited number of trucks.

I flew a Mosquito down and there wasn't much room in a fighter bomber. But you didn't need much in Burma—a couple of pairs of shorts and a few shirts.

When the monsoon came we were bogged in mud. We just had to wait as we couldn't get airborne.

The Japs dug trenches across the strip before they left but the army filled them and firmed them down very well before we arrived. Off the strip you had to be very careful. The strip itself was marked with 44-gallon drums and you wouldn't want to swing off.

One of the first Mosquitoes to be attached to 27 Squadron, Agartala India, April 1943.

Burmese pagodas—landmarks for pilots.

Kinmagon Airfield, Burma.

I did swing off once. There was a bang over the target and I didn't know what it was. Everything seemed to work all right. The undercarriage came down when I came in to land but a tyre had been shot off when a shell burst underneath it.

I landed on one wheel and kept it straight until I felt the speed dropping off. I thought we'd got away with it but I slewed off the runway because I burnt out the port brake. I steered between two of the drums and the port brake seized just as the starboard wheel fell into one of the trenches and we just flipped upside down.

Signals to land were given by Aldis lamp from the ground. We had a platform about ten feet from the ground to act as a sort of control tower.

A lot of our trips were one-offs. We'd be called to a village where they had oil dumps, for example. There are a lot of pagodas in Burma and they were generally very good landmarks. We'd get a direction such as 'a few miles north of the pagoda' and we'd go and do a few strafing runs.

On other sorties we were controlled by visual control posts of the 14th Army. They were in tanks waiting to advance and we'd go in with four 500-pound bombs. We glide-bombed in Mosquitoes. We could see the Japs on the perimeters perhaps 1000 yards ahead of the tanks.

The control was very good. They thanked us very much and we would see the British tanks going in after our attack.

We had one unfortunate case of bad luck and good luck down near Mandalay. One of our fellows was hit badly and had his arm shot off. He got the aircraft up to a respectable height and he and his navigator baled out.

The navigator landed all right but the pilot got hung up in a tree and bled to death.

Some friendly Burmese picked up the navigator. They passed him from one to another and they got him back to the squadron. It took some five weeks to get back but they looked after him pretty well. He was sent home.

We couldn't get motor vehicles in with supplies. The only road through to Assam was tortuous. It had taken our advance party three weeks to get through. The army had to rebuild the road through the Chinn Hills and repair washaways. It was very difficult to get supplies down to the plain where we were.

Petrol came in Skymasters and DC-3s. It was in 44-gallon drums and was unloaded by rolling them to the door and kicking them out, one on top of another.

Our bombs came the same way, all lying in the bottom of an aircraft. The crews stood and booted them out. I can still see the bombs dropping on top of one another—it used to give me the horrors.

Then the bombs would lie out in the hot sun—and it was very hot. You could barely touch them. The ground staff had to wear thick gloves to handle them.

The pilots had to supervise the bombing up of their aircraft and the 'erks' would always be having us on. They would be counting the number of turns of the fuse, 'Was that thirty-four or thirty-five or what?' I wanted to start running.

We lived in tents. I took a camp stretcher which was a real luxury.

I still have my old canvas bath. It is a bit of a frame with the bottom of the bath itself sitting on the ground. We used to sit there with just a bucket of water—water was scarce and undrinkable. There would be rows of men having baths outside their tents, especially after a game of volleyball. Volleyball was one of the few leisure activities. We played soccer and rugby but the ground was very very hard.

Canned M&V [meat and vegetables] was the main food. We had baked beans for breakfast every day—not the nice soft ones we know here, but hard white ones. We tried everything to soften them. I'm twelve stone now and not fat. I was ten stone then—we sweated it out and fined down.

Everybody got prickly heat there—especially the English. To alleviate it we were ordered to wear only shorts—no shirt—with boots and socks and badges of rank on our wrists. We had to send some people back home because of prickly heat.

I had a bearer—an Indian. I smuggled him into the base down at the feet of the navigator in the Mosquito. Sammy was a very good cook. We went out shooting hares and birds whenever we had the opportunity and Sammy would make curries—I don't know how. It was the only relief from M&V.

We got Sammy into Burma. The navigator said, 'You'll be in the can when the CO finds out.' The CO said, one day, 'You know, there's one thing you've forgotten, Shearer.' (I was called that because I was the only Australian on the squadron.) I said, 'What's that?' He said, 'We haven't got a bearer.' I said, 'Yes we have.' 'Where did you get him?' 'I brought Sammy down.' He said, 'You can't do that—he has to go back to India.' I said, 'It's a long way to go.' He stroked his chin, 'Well, we'll think nothing of it if he acts as bearer for me too.' I felt you had to take the initiative.

Sickness took a heavy toll throughout the Burma campaign. Malaria in particular was a major problem. The use of the drug Mepacrine and the spraying of DDT helped to keep the disease in check, but people were still infected.

Bill Hughes was one:

I got malaria mainly, in my view, because we had been doing an epic number of trips. On any trip with an afternoon take-off you'd land at an advanced base and come back the next morning when the cloud had dispersed. You slept in the aircraft—there was no accommodation.

You didn't get much sleep. It was stinking hot and there was a choice of closing up the aircraft and suffocating or letting in the mosquitoes. I reckon that's how I caught it.

I had terrific pains in the stomach. The squadron doctor said, 'You've got acute appendicitis.' There were two other squadrons in the wing and their doctors were consulted. They agreed. The nearest hospital to Kinmagon was at Myingyan, on the Irrawaddy. It was about forty miles away so the doctor said, 'You'd better go into the field hospital.'

They put me on the back of the fifteen hundred-weight truck and two or three other pilots went with me. I've never seen anything like the road. It's a wonder we didn't die in the process.

The hospital building had been bombed out and it just had a canvas roof but no windows or doors.

The doctor was a major. He looked at me and said, 'Yes, you've got acute appendicitis. I'm going to operate on you tonight.' I said, 'Now wait a minute. What kind of equipment have you got?' He looked at me as if to say, 'Who does this fellow think he is?'

I had visions of him having a jam tin which he had sharpened and was going to operate with. 'Have you got all the gear?' I asked. He said, 'I think we have enough.' I said, 'I suggest you leave it where it is.' He said, 'You might die during the night.' But I still resisted so he said, 'If that's the way you want it, be it ever on your head.'

I was propped up in bed and had a pretty hectic night but got a little bit of sleep in the early part of the morning.

When I woke up there was an army captain in the bed next to me. He had a pile of blankets but continued to shiver. I said, 'What's your problem? Have you got malaria?' He said, 'No, I think I've got dengue fever.'

They took blood tests from both of us. The major came back and said to me, 'You've got malaria.' I said, 'You told me last night I had acute appendicitis. Now you tell me I've got malaria. What's the chap over there got? Has he got malaria?' 'No, he hasn't.' So I said, 'Are you sure the Indian orderly that took the samples hasn't mucked them up?'

They decided on another blood test. The major came back and said, 'You're lucky to be alive you know. You've got malaria, the other fellow has dengue fever.' I observed, 'Thank God you didn't operate on me. I'd be a cot case now.'

I was very sick for a week and remained in the hospital. It was a terrible place.

When I got back to the squadron the CO said, 'You're going away on leave.' I said, 'It'll take me a week to get out of this place. Why don't I go shooting with my navigator?'

On one of these shooting trips (we were shooting pigeon and even little doves—anything we could eat) we came round a clump of bushes and there was a Jap. He had a rifle and we both had shot guns. He turned one way and we turned the other. We only had bird shot and didn't want to engage him. We didn't know whether or not he was a scout. He was the only Jap I ever saw at close quarters.

Rangoon

On 28 March 1945 the Japanese forces began to retire from Meiktila and the way was open to Rangoon if the monsoon rains did not prevent it.

The race was on between XV Corps coming from the Arakan and IV Corps advancing towards Toungoo, Pegu and then Rangoon. Air activity over Toungoo and surrounding areas was intense.

Cowan's 17th Division reached Pegu, fifty miles from Rangoon, just as the rains came and halted the advance. At the same time, on 28 August 1945, XXXIII Corps took Allanmayo and proceeded to take Prome.

Bill Hughes checked whether Allanmayo was still occupied by the Japanese:

I did one trip because the 14th Army wanted to know whether the Japs had pulled out of Allanmayo. I was to go down and do a recce. The idea was to draw fire at the aerodrome to see if they were still there.

I flew across a corner of the aerodrome while flying south. It was all quiet and my navigator said, 'It looks as though they've all gone.' But I said, 'We'll do a recce further down the Irrawaddy.' Which we did.

About half an hour later I came back and crossed the drome at a different point. The heavens opened up with Bofors and what not and I was ducking in the cockpit—it was quite instinctive.

I was at zero feet. I opened up to full bore and I could see a glow in the starboard engine out of the corner of my eye. I wasn't game to look—I thought it was on fire. I had my head down and stuff was going everywhere. We got out of range and everything became peaceful again.

I looked out and saw that the whole of the engine cowling had disappeared from the starboard engine. All I could see was the red-hot cylinders [probably the exhaust stacks as the Merlin engines were water cooled]. I thought, 'My God, I've still got full power on,' so I pulled the power off and the glow disappeared.

We got back and taxied into the dispersal bay. In a flash one of the airmen was up on the mainplane. He had used a stepladder and come up through a hole in the wing. The outer tanks had been blasted away.

It didn't affect the handling although the plane did appear a bit sluggish on the way home. Both engines were still working.

Bill had another 'shaky-do' just north of Rangoon:

I once told my navigator to bale out. He refused to go, saying he wasn't going to jump into that country. Apart from the jungle the Japs were the big problem. They were staking fellows out in the sun. They didn't like the air force.

I was leading four aircraft and lost my port motor after I left the target. I landed at an advanced airstrip which had just been opened a little north of Rangoon. The army had just taken it over from the Japs.

The hydraulics had gone as well. I had one leg down and no flaps. I thought the other leg might make it but it didn't. I came in at 170 mph and landed on one wheel. The strip was all-weather 2000-yards and I did a glorious ground loop at the end. The aircraft was a write-off.

We had just landed and got our gear out when an army major came over in a jeep. He said, 'Welcome. You're the first aircraft in. There are to be big celebrations and you'd better join us. We have had a hell of a battle. We often lost the place at night and had to get it back the next day.' I thought, 'I'm not staying here.'

The major said, 'Come back to the tent. You might like some breakfast.' It had been a dawn raid and was only about eight o'clock then. They brought out bacon and eggs. We'd never seen them in Burma.

I said, 'Where did you get this from? We don't get this. You're living on the fat of the land.' He said, 'Oh well, special occasion. Big celebration,' and gave no more information.

I used his field telephone to report to Field HQ and said we wanted to get back The AOC sent his own aircraft to take us back to the squadron.

Bill Hughes was over Rangoon the day it was occupied:

We went down when the invasion took place. We had bombed Battery Point, at the entrance to Rangoon, the previous day. There was a big installation of guns there. The squadron put up twelve aircraft with four 500-pound bombs each and we just went in.

On the day of the invasion we flew over and saw the message laid out on the rooftop of the prisoner-of-war camp. It said, 'Japs gone. Extract digit.'

We knew no Jap would say 'Extract digit,' so we reckoned that it was for real.

At this stage an air/sea invasion by British forces was imminent. There had been rumours for months. Ron Gardner recalls one:

To provide close air cover for the invasion of Rangoon there was a proposal to load Spitfires on to aircraft carriers in India and sail along the west coast of Burma. The flaps on a Spitfire had only two positions - fully raised or fully lowered - so, to give enough lift to take off from the carriers, it was proposed to raise the flaps with chocks of wood inserted above them to hold them lowered at about 25 degrees. Once airborne, the pilot would lower and then raise the flaps to allow the chocks to drop out and continue on the mission. Fortunately, they captured the airfields early and didn't need to try this.

Doug Nicol was with the forces about to launch a seaborne assault. He was flying in Catalinas at the time, and took part in the operation. This is how he recalls it:

In the invasion of Rangoon, we were stationed on a ship, the Manoora, which was moored in Akyab Harbour on what was really D-Day. The planes were moored around the ship, and we fuelled from the ship. We had a base at sea. It was a quite unusual situation.

We were given two cans of beer on the eve of the operation. We hadn't seen any since we left the UK. It was a pretty weak brew, but some of the boys didn't drink and gave theirs to the drinkers. There were some sick boys next day, which was the day they had to take off for the invasion.

Our Catalinas were to provide a patrol to ensure that the Japanese subs and aircraft didn't attack the vessels engaged in the landing. We had to rely on visual sighting for any action or warning against subs—a daunting enough task, and the Catalina couldn't provide any defence against a determined fighter attack.

In fact, the Japanese had left Rangoon. The Allied forces landed with few casualties (some, unfortunately due to Allied bombers dropping their loads on some of their own troops).

Rangoon was occupied by British troops on 3 May 1945.

The Final Phase

The fall of Rangoon did not signal the end of the war in Burma.

The Japanese still held pockets, some of them quite large, spread out over a large area. They were congregated outside the north-to-south corridors from Toungoo to Pegu and from Yenangyaung to south of Prome and the west-to-east corridors from Meiktila to Taunggi and from Toungoo to Mawchi.

Slim issued orders to advance on Moulmein, take Mokpalin, destroy all enemy forces in the Irrawaddy Valley and open the road from Prome to Rangoon.

John Eather was flying at this time:

We got conditioned to flying at 14–15 000 feet without oxygen.

A typical sortie was from Ramree Island (south of Akyab) to Toungoo which was a main supply base. A clear run took three hours—one hour there, unload, an hour back, reload, etc. Sometimes we'd do three trips a day. We'd take off as early as we could because the weather got worse as the day went on.

The cloud was in layers. Once through the first layer of several hundred feet you would fly in a sort of cave of cloud with a cu-nim coming up every now and then.

I wondered what would happen if I flew into one of these unseen. An old pilot gave me a tip. He said that if the radio compass was tuned to an unused frequency the needle would point to a disturbance. I tried it in clear air by flying round a cu-nim. It did roughly follow the disturbance.

I had many trips to Toungoo. At the same time we were supplying down the Arakan and the Irrawaddy. The Pegu Yomas was the area between these—it was considered the worst malarial area in the world.

When I had 400 operational hours up the tour duration was changed to 700 hours. My diary entries record the declining balance remaining. At nine hours I started to get edgy.

At he beginning of May my diary reads: 'Peace in Europe. Offerings all week. At last it came. Today Churchill spoke. Very dramatic. All the mob getting drunk. I'm on first sortie tomorrow so I'll have to hit the squash. Let's hope this doesn't take long. Have been down as far as Rangoon. Ready to visit Ramree Island this week. Should be OK. Hope to be going to Rangoon.

There's lots of fuss about peace in Europe but we're still here. The forgotten army and the forgotten air force.'

Htuak Kyuat War Cemetery Rangoon. The resting place of some Australian aircrew who died in Burma.

A typical headstone.

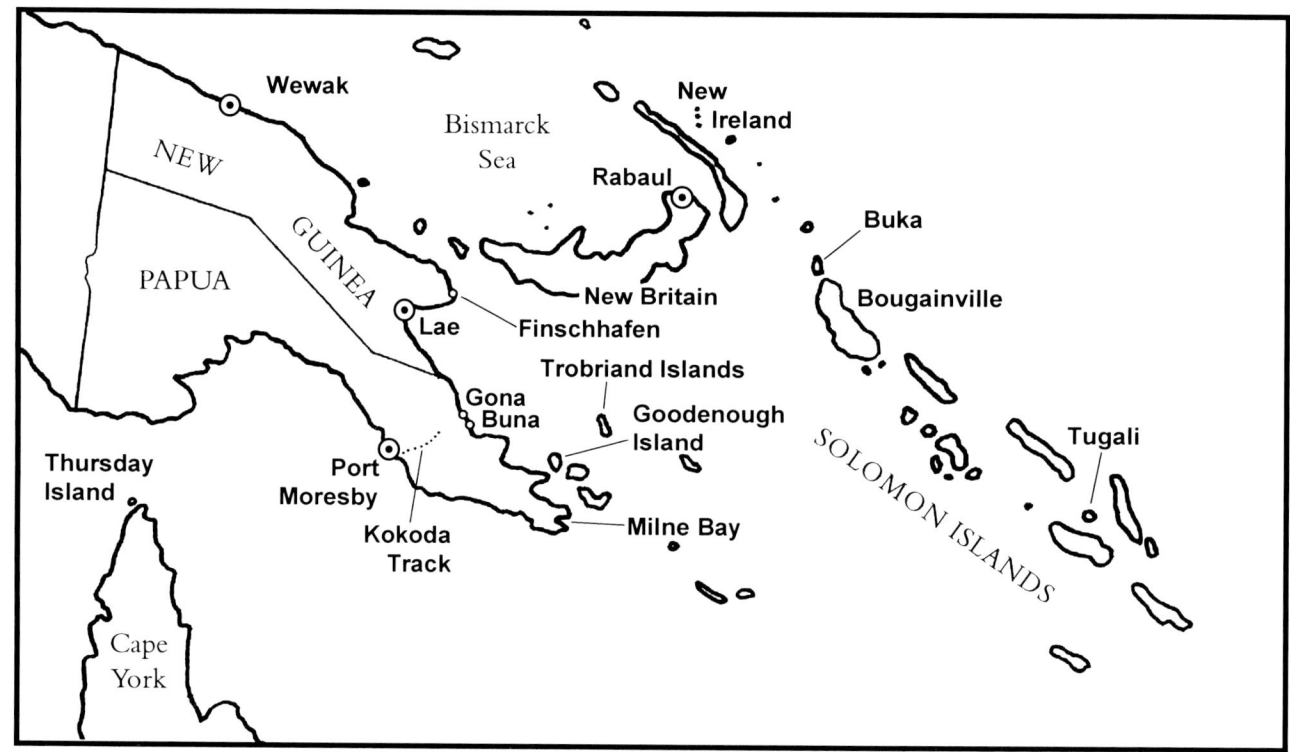

SOUTH-WEST PACIFIC THEATRE WWII

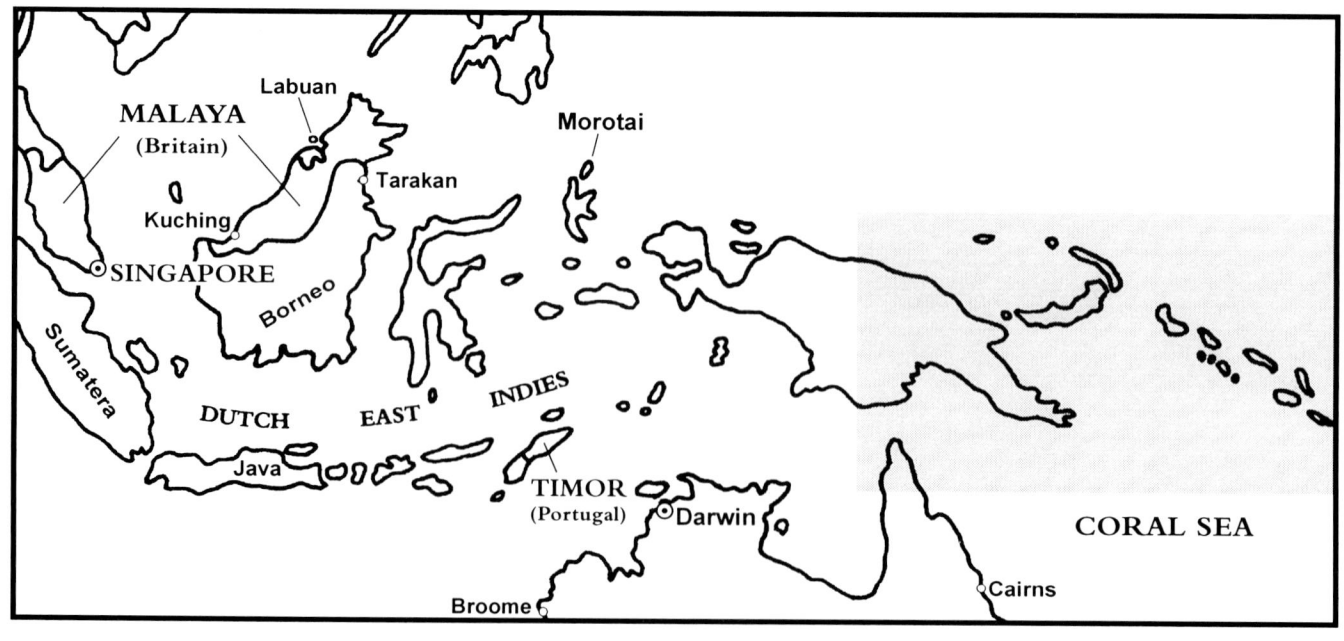

4

THE SOUTH-WEST PACIFIC

AUSTRALIA

WESTERN AUSTRALIA

Western Australia is a state of vast distances. It borders on the Indian Ocean, across which our troopships went to the Middle East. This ocean was also the route for supplies to Britain.

The ocean was patrolled by the Royal Navy, to which was attached HMAS Sydney, of the Royal Australian Navy.

• The Loss of the HMAS *Sydney*

The RAN cruiser Sydney had a distinguished record in the Mediterranean, where she served from May 1940 until January 1941. During this service she sank the Italian cruiser Bartolomeo Colleoni and the destroyer Espero.

In January 1941 she came home and carried out her duties in the Indian and Pacific Oceans. On 30 November the Prime Minister, John Curtin, announced that she was presumed lost after she had sunk the German raider Kormoran.

Air and sea searches were carried out, but to no avail.

Eric Cooper tells of his role:

The *Sydney* had been into Geraldton a short time before she disappeared, and we had had a lot of the officers and crew in the mess.

When she disappeared, we searched from Geraldton to Carnarvon and out to sea. Five or six aircraft participated. We flew out of sight of land, on a predetermined course and landed at Geraldton.

We didn't sight anything on the flight, but near Carnarvon, we did see one of the ships which had come in with some of the German survivors in tow in a lifeboat.

We were using training aircraft which were being used for pilot training. This involved training in aircraft handling—circuits and bumps, normal take-offs and limited cross-country flights. Consequently, they were not fitted with radio or accurate drift sights.

We could only use dead reckoning on the search for the Sydney. In fact, the winds increased while going out from Carnarvon towards Onslow and we finished up some miles north of Onslow. We had to track down by map-reading along the coast. In those days, the coast looked pretty featureless—it was virtually uninhabited.

We were to carry out the search from Carnarvon to Onslow, and arrived there late in the evening. One of the flight didn't arrive, so we had to arrange a search from Air Force HQ in Perth.

Perth said an aircraft would be coming up first thing in the morning but, in the meantime, an aircraft equipped with radio had joined the search and the pilot reported to Onslow that he had sighted the missing aircraft on a mudflat at the bottom of Exmouth Gulf. He said he would land beside it.

He subsequently reported that he had landed all right, but had bogged his aircraft. So, now we were two planes down. The first was out of fuel, the second could fly, but was bogged. They needed another aircraft to bring some fuel and shovels, etc., and then they could all get out.

I loaded my Anson with cans of fuel and a few shovels, flew to the position they had indicated and landed on the spot they had marked out as being firm and safe to land on. Unfortunately, just at the end of my landing run, my aircraft broke through the mud crust. One side was buried in mud up to the engine. Another Anson gone. We walked straight out of the plane on to the ground.

We couldn't get the aircraft off, as the power of the engines was not sufficient. We reported the situation back to Onslow, who contacted the local sheep stations. One of the managers came out, across the mud, to our aircraft, at about 10 p.m. We were about a mile from firm ground.

We spent the night at the sheep station. We finally had to get a ground party up from Geraldton to get the aircraft out.

• Broome

The north-west coast of Australia, as well as the Northern Territory, Darwin in particular, were the subject of enemy action.

The Japanese attacked Broome on 3 March 1942, causing considerable damage to the harbour.

One witness was Keith Caldwell, who was a Flight Lieutenant flying boat captain:

I was in Broome at the time of the first attack by the Japanese. Group Captain Scherger asked me if I would go to Timor and pick up about fifty people who were stranded there by the Japanese troop movements. So I said, 'Yes,' and said, 'Where do we go from? Broome or Darwin?'

I thought I would rather go from Broome, because we might have better winds and use less fuel. So it was arranged that we refuel at Broome and take off. Unfortunately, Lester Brain had grabbed all the fuel for Qantas and there was no fuel for me for a few days. Even by the third day, we hadn't been able to get our own air force, which was assisting the Americans, to do anything for us. Our messages didn't seem to help us to get fuel, so all we could do was wait. We were there three days.

As it happened, on the day of the raid I had gone ashore with my radio operator. The rest of the crew were able to handle anything that needed doing on the aircraft, so we went to the control tower. Just as we arrived—it must have been about a quarter to seven in the morning—the lookout (an American officer) called out from the top of the tower 'I don't know whether it's of any interest to you Australians, but an aeroplane flew over about half past two this morning. We challenged him, but he gave the wrong reply. We challenged him again, and again he gave the wrong reply. He flew round for a while, but he's gone now and we haven't heard any more about it'.

The timber I had ordered had arrived, so I loaded this and we headed back down the road. The tower was about a mile and a half to two miles from the aircraft. My radio op and I were both standing on the running boards—he on one side, and me on the other.

Out of the corner of my eye, I saw aircraft attacking at treetop height—about twenty or thirty feet up—and coming right past us. I told the driver to go as quickly as he could. But there was no way we could keep to the road—we would have to go into the bush and try for some sort of cover.

We arrived at the cliff tops to find my aircraft in flames. The others were soon afterwards.

There were fifteen seaplanes and fourteen landplanes there. They had all loaded up, with their passengers, waiting to take off. All the passengers—over 200 of them—were killed. My crew was the only one to get out.

My radio op raced to our plane to see what he could do. He was standing, with his legs apart, on the plane, alongside a 44-gallon drum when bullets went between his legs. They set the drum on fire—he was lucky he didn't get burnt. He jumped over two bulkheads, ran up, on to the flight deck, right alongside the three internal tanks, which went up in flames almost immediately. He picked up a rubber dinghy and took it back down the companionway to the lower deck. He then pushed it through the front door. As he did so, he bumped the button and the dinghy went 'whish'—he thought it was another tank going up. He picked the rest of the crew up out of the water. All the crew were in the dinghy and got ashore safely, where they began helping me pick up the bodies. They were very badly burnt bodies from all the other aircraft. They had gone up in flames with their fuel tanks full and their passengers inside.

The Japs flew down the long jetty—about a mile long, I think. Instead of just going up and down it and shooting everybody, they made a ninety degree turn and just shot off a few rounds to set fire to some of the drums on the jetty. They didn't go for the people.

Keith Caldwell goes on to explain how he 'escaped' from Broome:

My crew, except the radio operator and myself, were flown to Port Hedland for transport to Melbourne and Sydney, etc.

My senior radio operator stayed in Broome with me and, cautiously, a couple of days later, we joined the lines of aircrew survivors waiting to board evacuation aircraft. One of these American aircraft crashed in the haste to take off. No wonder! Some of the aircrew had been flying without rest for many days and nights. I recognised an American captain, and was offered two seats, for my radio operator and myself.

Unfortunately, his navigator was inexperienced and incapable. So, uninvited, I took over the navigation to ensure we would really get to Melbourne.

It was just as well. After we crossed the South Australian border the aircraft turned north-east, towards Canberra. I protested that we could not make it, and persuaded them back to the original course, after corrections, to Melbourne.

The two outboard engines cut as we were landing, and a couple of minutes later the two inboards failed. We missed the fence by a few inches (less than ten) and were towed by a tractor to a hangar.

That crew were then sent to Darwin. Much less than twenty-four hours later, all lost their lives in a crash just after their Darwin take-off.

THE NORTHERN TERRITORY

• DARWIN

Situated at the northern extremity of Australia, Darwin was very vulnerable to attack when the Japanese started their thrust south in 1942.

It was the first Australian city to be attacked when the Japanese launched two raids on 19 February 1942. The attacking force was large: seventy-one dive bombers and eighty-one level bombers launched from carriers with fifty-four land based bombers.

The harbour was crowded with forty-five ships. Five of these were sunk and a number damaged.

The RAAF base was unprepared and confused, with about twenty aircraft being destroyed there as a result.

A significant number of civilians were included in the 250 people killed.

Bob Crawford, a Wirraway pilot, has this to say:
The first direct action I saw was on 19 February 1942 when Darwin was bombed by the Japanese—the first air raid on Australia.

Our flight was on the civil aerodrome at Darwin. It was about two miles as the crow flies from the RAAF aerodrome. The rest of the squadron was at Batchelor, fifty to sixty miles south of Darwin. The squadron comprised of six Wirraways which were a general-purpose aircraft.

About nine o'clock in the morning of the big raid my flight commander came along and said, 'Crawford. Go and get your navigator. We're going to go flying.' I said, 'What for?' He said, 'We've got to defend the ships in the harbour.'

My navigator was doing the washing. I said, 'We've got to go.' We were walking down between the huts when the flight commander ran past us. I said, 'I'm sorry sir, I haven't got the aircraft out.' He said, 'Bugger the aircraft! Look!' We looked up and there was the raid coming in.

We ran into the bushes where our aircraft were dispersed and spent the next hour and a half to two hours putting out fires on the aerodrome or in our huts.

We couldn't do anything to drive off the raid as the bombers which attacked us were above 20 000 feet, well above our operating height of 6000 feet, and we were

Wirraway aircraft.

still walking to our aircraft when the Zeros started strafing at ground level, which made it impossible to take off.

Then we were ordered to go over to the RAAF aerodrome and get some armour-piercing bombs which were dispersed in the bush over there. We were hampered in getting there by the number of people on the road evacuating Darwin and going south.

We collected and brought back the bombs which the armourers then put on the planes (four 250-pound bombs on a plane). We waited for further orders. Finally we were told to stand by until the Hudsons of 13 Squadron returned. They had gone out to locate the Japanese fleet which was assumed to be about to land troops in Darwin. We remained on stand-by throughout the second raid.

Once it got dark it was impossible to fly so we were stood down.

We didn't lose any aircraft at our drome—they were all dispersed and camouflaged. There were American P-40s at the RAAF drome and some had taken off on their way to Timor when the raid came in. Some got into combat and some were destroyed on the ground.

We were on alert all the next day. I had to do a search for any signs of an invasion force. I flew 100 miles south-west from Darwin and then flew in an arc which brought me round on to the northern tip of Bathurst Island. My orders were to radio back the location of any force and then attack. The rest of the squadron joined in this surveillance after my initial search.

There was a lot of talk of a fifth column in Darwin and it was rumoured that people were leaving sails pointing in the direction of Darwin airport. So when we went walking at night we would turn anything that pointed toward the aerodrome so that it pointed in another direction.

Our work revolved around looking for scattered ships. I was sent out a few days after the attack to look for an American ship, it was 3000 to 5000 tons, which had lost contact. I found it burning on the west coast of Bathurst Island.

There were some bodies of dead servicemen on the ground but I couldn't land because the sand was too soft. I couldn't raise base with the 1082/1083 receiver/transmitter equipment so I had to get back to base before I could report it.

These searches went on for about fourteen days—we navigated by map-reading and DR. There were a few raids on Darwin during this time.

Then I was suddenly posted to 75 Squadron.

• The Milk Run to Batchelor

Bruce Edenborough was a pilot on the milk run:
I was one of the foundation members of 36 Squadron. When I look back I realise that the state of Australia's preparedness for war was rudimentary.

36 Squadron operated from Laverton with a collection of different types of aircraft which had been hurriedly collected to make a transport squadron. It had a Tiger Moth, a couple of old Junkers from Papua/New Guinea with their corrugated sides and open cockpit, DH-84s and DH-86s. The pride of the squadron were the Douglas DC-2s. They came from Eastern Airlines in America. They already had many hours up and were old planes.

I was a second pilot for most of my time with the squadron. The first pilots were nearly all fairly experienced civil pilots. Our tasks were many and varied.

We used to call the trips to Batchelor 'milk runs.' We would fly up through the centre to Alice Springs, on to Daly Waters and then to Batchelor. There were times when we carried two aircraft engines in the passenger section. They would be unloaded at Batchelor and replaced by two other engines to go south for overhaul. While they were strapped down securely, we had no illusions as to what would happen if we crashed with them behind us.

At Batchelor we had our first experience of multiple toilets. Here you would sit, about twelve in a row, with no partitions and everyone watching to see that you didn't use more than two sheets of toilet paper. There was a terrible shortage at the time.

We heard about the raid on Darwin while at Batchelor but did not experience it ourselves. We were told of the primitive conditions in Darwin after the bombing.

I can remember someone telling me they had run out of glasses in the pub and you had to bring a tin or a pannikin to get a drink. It was claimed that some even offered their hats to be filled with beer and one bloke was so thirsty he offered his boot which was duly filled and the contents enjoyed by the thirsty one.

Queensland

With the entry of Japan into the war in December 1941, north Queensland, in particular, became an important staging area for forces en route to New Guinea, either by ship or air.

Keith Caldwell was stationed there in May 1942:
The most difficult of all my operations in flying boats took place on 22 May 1942. I was to take Captain Herd, of the Salvage Board, from Townsville to False Orfordness, on the Cape York Peninsular, so that he could make an attempt to salvage the corvette *Ballarat*. I found the vessel just a few hundred yards north of 11 deg. 23 min. S., and 143 deg. 53 min. E. long.

Conditions at False Orfordness were very much worse than anything I had ever handled before.

The corvette was high and dry on the sandy beach. Its propellers were completely out of the water. Very big seas were rolling in from the ESE and they had been turned by the headland from the SE, so that they were somewhat erratic after they came round that rocky outcrop. I made a full circuit to observe them.

Getting in would not be impossible. Getting out might be. Obviously, the corvette was more valuable to Australia than the flying boat at this time. HMAS *Townsville* was standing by to assist with the salvage.

I decided to slide down the steep hillside, with all engine throttles closed and flaps full out, just above the stall speed, and, hopefully, to touch the keel in the wet sand at the water's edge. I was very lucky, everything went exactly as planned. The speed dropped off rapidly. My crew were all in their correct positions for alighting and anchoring before we hit the very big waves. It was impossible to stand without holding on. The motion was violent and irregular in the swell and cross seas.

Two anchors were dropped on the move so that, as the flying boat dropped back, it would ride at the end of a 60 degree triangle. Then all four engines cut. Now the motion was really violent. It was difficult for the crew to hold their planned positions and carry out their duties.

The pinnace was signalled to come alongside so that Captain Herd could be transferred. The coxswain proved to be a miracle man. Our boat was swinging through something like 60 degrees, in an irregular motion, and rising and falling at a completely different time to that of the pinnace. The coxswain held the bow of the pinnace a few inches clear of the trailing edge of the wing, and the mast was only just clear of the leading edge of the tail plane. There were only a few inches between the sides of the pinnace and the flying boat. The most peculiar thing of all was the fact that, for much of the time, our big promenade door was wide open and two or three feet below the level of water in which we and the pinnace were riding, but no water came in.

Actually, the transfer operation was completed in less time than it takes to read the story.

Captain Herd was standing near the door, ready to jump through as the pinnace went down and we went up. The difference in the rise and fall was due to the difference in our water line lengths and the staggered position of the two boats. Our team of lookouts, watching the seas ahead, the swing of the boat, the size of the wave, etc., passed their movement information until 'GO' was given. It was somewhat frightening to watch, but the transfer was made in less than three seconds. It must have been a frighteningly long three seconds for Captain Herd.

Now it was our turn to try and take off. In our position, we were sheltered from the really big seas but, even though we intended to drift backwards after picking up our anchors, we would be in the really big seas before we had started to gain speed.

We watched the waves ahead from the top of the hull while we were still anchored. I had gained a good idea of what we were up against. I judged that the pattern of the waves was, if I remember correctly, three moderate waves, fairly close together, then about 180 feet to a monster wave with a very steep front, having a difference in height, from top to bottom of the long trough, of about twenty feet. That was twice the height of the flying boat's hull above the water line. Then there was another pair of big waves at a similar distance, another gap, and the pattern repeated itself. The waves were steep-fronted and, as indicated previously, there were also the not insignificant cross seas from the south east.

With anchors up, all four engines running and still warm, I allowed the wind to take us back as close to the shore as possible. Then, calling, 'Stand by for take off,' I opened the four throttles for maximum power.

Although the craft responded well, in just a few seconds I found my responsive reaction to be working faster than normal brain processes. I knew that I must, even at the lower speeds, bounce the aircraft so that it would never hit a wave front. So long as I could slide down on to the top, or the other side of the top, I should be able to time and create bounces or simply gain speed, to hopefully keep me out of trouble, as we dropped to the back of the wave. A tiny bit of rudder, judiciously applied for a moment, would thrust the boat slightly upwards and help the speed gain. This was a process of applying skills learned earlier in my career, when operating my Cutty Sark amphibian.

I will never forget that take-off, as long as I live. Instead of the fairly normal time of thirty to forty seconds, in calm water, or about sixty-five to seventy seconds in an overloaded boat, it took two minutes and twelve seconds. We were airborne and safe.

Fred Cassidy had a narrow escape at Townsville:

When we were at a drome just east of Townsville, we were told, 'You'd better do some sea searches and night flying to kill time while you're waiting.'

We hadn't done much night flying, so we went out to sea from Garbutt and came back to land. We were half-way down the flare path, at about thirty to forty feet, when the lights went out. We banged and bumped and eventually got the thing down, but it was a shocker.

After we got out, we went to the Duty Pilot's tent and asked what the Hell was going on. He said that there was a truck coming along with its lights on and he yelled, 'Turn out those lights!' The airfield crew heard him and turned out the flare path lights. You don't need enemy action to kill you.

Bruce Edenborough tells of 36 Squadron's sojourn at Townsville:

The squadron moved to Townsville and to the Stockroute Strip near Garbutt. We were involved in biscuit bombing on the Kokoda track in the hills behind Port Moresby.

After I had been with the squadron about twelve months every DC-2 had ended its life.

One was caught in a cross-wind at Cooktown and crashed; another crashed on the coast at Mallacoota; another crashed at Moresby. We had to fly into Port Moresby at dusk because of the danger of being shot down. This poor chap was coming into Seven Mile Strip. As it was wartime they did not light the tops of the hills—the strip was not lit too well either—and he let down too early. He crashed into a hill and everybody on board was lost.

On one occasion we set out to come back to Garbutt field at Townsville and made a landfall about halfway between Townsville and Cape York. Because of cloud we could not make out where we were and just had to follow the coast down. We had set our course on the winds given by the Met people but they were wrong and we were hundreds of miles off course.

The DC-2s were not nearly as stable as the DC-3s. The DC-3s had better rudder control in the air and were more stable in a cross-wind on the ground. This was generally not a problem but Cooktown was a bad strip. It seemed to be laid so that there was always a sea breeze blowing right across it.

There were a number of strips at Iron Range. The Americans operated from there at the time of the Coral Sea battle.

On one of our trips there we had to leave the plane parked on the runway overnight. In the morning we found that a jeep had run into a wing and damaged it. This threw things out for a long time as we had to wait while it was repaired.

I remember they were showing a western film at this strip one night. Just at the critical moment, when the stage coach was going through the canyon and about to be attacked by Red Indians, someone in the audience jumped up. Then someone shouted, 'Look out!' and others jumped up. Then a voice yelled out 'Air attack!' and the whole mob got up in a rush to get out and trampled others still struggling to get to their feet.

We were told it looked like a battleground afterwards with people spreadeagled everywhere. All because someone thought he saw a snake.

Ron Barker was engaged on an unusual experiment in North Queensland:

After I came back to Queensland, I was sent to a chemical research unit. The unit had Vultee Vengeances at the time, which was probably why I was sent there. We carried out the bombing side of war gas experiments.

The experimental station was at Proserpine, just near Mackay, but the aerial side of it was carried out from the civil aerodrome at Bowen, where we had the Vultees. Our duty was to load up with gas canisters and drop them in certain allocated areas.

One area was at Tully, where there was a dense rainforest. The experiment there was to find out how long it would take for liquid mustard gas to dissipate if it was sprayed in the jungle. That would have been handy to know if we had been attacked by the Japanese in jungle areas. I believe that, even today, there is a danger that some of that gas exists—even after fifty years.

The main experiments were carried out at Proserpine, where volunteers were in trenches and dugouts. These volunteers were from the three services' and were paid an extra sixpence a day for their efforts. We would fly over and drop the gas canisters on them. The canisters would burst and the area would be covered with gas. They would be fully protected with clothing and respirators. The experiment was to evaluate the protective clothing.

Bruce Edenborough adds to Ron's story:

We had to mount sorties on troops in the jungle. We would fly to Cairns where we would be given the final details of where the soldiers would be. We would fly there and carry out our bombing or spraying on them. The destinations were some of the tourist areas of north Queensland. Mission Beach was a place we visited many times. I have a record that we even went to Green Island and other islands off the coast which were used because they had thick jungle.

On one trip we flew in line astern over the troops on the beach. This was dangerous because we ran into each other's spray. The first I knew about it was when our plane was absolutely spattered with mustard gas— great particles of green and yellow mustard gas. The plane was immediately filled with the smell of mustard. We came home very worried and wondering whether we had been affected. We did have sore throats but the plane was, fortunately, put together well enough to not allow more than a sniff to penetrate.

A lot of the Australian soldiers were very badly burnt. We went to a hospital in Proserpine and saw them with terrible red weeping blisters. They were made to carry on as long as they could. They ended up just able to stagger round with their feet wide apart because their crotches were full of weeping blisters. The experiment was to find out how the gas would immobilise the troops.

DC-3 Dakota in post war livery.

Passengers and crew of Ron Barker's plane with anti-mustard gas protective clothing.

Ron Barker has further comments:

When the war ended, we had quite a stock of gas to dispose of. It could not be disposed of over land, so we took it out to sea, in barges, and dropped it over the side. However, the drums floated, so we had an aircraft pickaxe with which we put a hole in the canisters so they would sink. But in putting the holes in them, the gas would spill out and spray. If you weren't careful you'd be splashed by it. I remember the warrant officer in charge of the party was sprayed, and a few days later he had huge blisters on his arms. They were the size of cricket balls.

We carried out one experiment in the air. London wanted to know what would happen if the enemy attacked an aerodrome with mustard gas. They particularly wanted to know what would happen to the aircraft. They took a Beaufort aircraft, sprayed it with mustard gas and called for volunteers to man and fly it for two and a half hours. Another chap and I volunteered, and we took two scientists with us—one of them was a WAAF officer.

The four of us were dressed in protective clothing and respirators, even while flying. The scientists carried out tests inside the aircraft. I kept constant contact with the base by radio and kept them informed of what was taking place. The whole flight was monitored.

We flew out to sea and up and down the coast for two and a half hours. We were at about 5000 feet and didn't know whether, if the engines suddenly stopped, we would be able to reach the beach.

When, after the flight, we came in to land, the brakes failed and we did a ground loop at the end of the runway. An examination showed that all the brake fittings had corroded.

I found that, toward the end of the operation, the gas concentration had built up in the radio cabin to such an extent that it was coming into the respirator and I was breathing in the gas. You could smell it.

The pilot had some difficulty controlling the aircraft as he came in to land because he was wearing heavy gloves and the goggles in his respirator were fogging up. So he took off his respirator and gloves. Both he and I, for some time afterwards, suffered from the effects of that gas. I was indisposed for a week, but he was in hospital for four or five weeks.

As a result, I suffered from cancer, had an operation and recovered. I still suffer from sinus problems—the medical staff found the gas had burnt out the protective tissue in the nose and throat. The woman scientist—the WAAF officer—suffered medically and died of cancer some years ago. Two of the four got cancer, and I'm sure it must have been because of exposure to gas.

NEW SOUTH WALES

• Nowra

When war broke out, Australia had few front-line aircraft. Its main planes were the Seagull (Walrus), the Hawker Demon and the Anson, all obsolescent. Nevertheless, the Ansons, as we note from Eric Cooper's comments, were pressed into service at Nowra, on the South Coast of New South Wales:

Ansons were a very kind aircraft—wonderful for pilot training. They had a slow landing speed and had no vices to speak of. Anybody could fly them. I flew them operationally for a short period.

There was an Anson squadron based at Nowra. It was used for coastal convoy work, from Gabo Island up to north of Newcastle, bringing ships up and down the coast. It was anti-submarine work.

I don't know what an Anson could have done against a submarine, but we carried a couple of depth charges. The aircraft were fairly slow, but they at least had the coast to follow. Their endurance was three to four hours.

As for the depth charges, they could have been quite effective if you got them close enough to a submarine.

As we had no radar, we operated only in daylight and relied on visual contact.

• Mascot/Richmond

Test flying was not the glamorous task depicted in some movies or books. It required careful checking and often became routine and boring and was often dangerous. Many test pilots longed to see action in a war zone but, because they were generally good pilots, they usually found it very hard to get released.

Bob Hartman was a test pilot at Mascot and Richmond and tells his story:

The test pilots were not always volunteers. I spent the first few months testing Tiger Moths at Mascot. There was an aircraft repair shop at Bankstown—Clyde Engineering. They did a lot of repairs for the RAAF on Tiger Moths, Wirraways and Ansons and I tested the planes from them.

At Mascot there was Ansett Airways controlled by none other than Reg Ansett himself. They were working for the RAAF on a cost-plus basis. They were working on a Fairey Battle and I was sent across to test it. I didn't have a lot of hours in Battles but I reckoned I knew more about it than Reg Ansett did. I flew it three or four times before I accepted it. Every time I knocked it back he'd do his lolly. I finally accepted it.

When we went to Richmond we were testing Spitfires and Kittyhawks. The Spitfires had been flown in the UK, boxed up, sent to Australia and reassembled.

The Kittyhawks had never been flown, simply crated and shipped.

I was in a Spitfire when the engine cut out on me about 10 000 feet over Prospect Reservoir. By the time I'd looked around for a handy paddock to land in and had gone through the cockpit drill the engine restarted. When I got back they found there was a lot of water in the system. If I had a load on the engine it cut out. If I put it into fine pitch for a landing the load was reduced and it restarted.

On another occasion the whole port-side cowling of a Spitfire blew off. I didn't know it had come off and couldn't understand why the aircraft stalled at a much higher speed than normal.

When you first go on a test flight you're told exactly what you have to do. Check the instruments, check the hydraulics and check the speed and then fill in a form showing the performance.

There was a test pilot's course at Laverton and I was sent on it after I'd been a test pilot for six months. It was a crash course but you learned the ins and outs of everything in an aircraft and what could go wrong with it.

They were always trying to get more trained test pilots about the place. Up in the islands, if there was something to be tested a squadron pilot would take it up, do three or four slow rolls, do half an hours aerobatics which he would have no chance to do otherwise and come back. The authorities were against this and wanted to get more specialised and experienced test pilots to do a complete job.

• Cootamundra/Evans Head

People tend to forget the part played by staff pilots. They were the men who were, in most cases, keen to see action but who were directed to return to fly and help train other trainee aircrew. Pilots and other aircrew had to perform these chores day in and day out. The planes they flew were not top line aircraft; they were old and tired. The men themselves were not old but were often tired, bored and disillusioned.

Bruce Edenborough can finish the story:

36 Squadron was told to shed some of its pilots. So right in the middle of winter a group of us were transferred to Cootamundra Flying Training School. We came down from the warm climate of Townsville to Cootamundra in the winter. I still remember how cold it was. My first act was to buy a sheepskin vest which I wore under my overalls.

Our task was to train navigators. We felt it was a bit of a comedown as we'd joined up to carry the fight to the enemy and not to be shut away in a training school. We didn't take it too well. We were browned off but did finally accept that this was part of the war effort. Navigators had to be trained.

Occasionally you would get lost on night flights. One reason for this was that lights in the towns were turned off at a certain time. This was partly to save electricity and partly as a war precaution.

We would do two trips on night flying. If your trainee navigator gave a wrong bearing on the first trip you could see where you were from the lights of the towns. But on the second trip there was just blackness and it was quite possible to get lost.

On the first trip one night I came back with a flat battery. I had no landing lights or radio—everything was out. The plane was refuelled and we went out again. We had a long exercise and coming back on the second leg the town lights were out and I didn't know where we were.

Then we saw lights in the distance. When I got down near them I realised it was Albury—a long way from Cootamundra. Then everything went black in the plane—no radio, no lights for the navigator to work by.

When I checked the petrol there was no movement and it showed zero. I felt we would not have enough petrol to get back. We descended and found the drome at Wagga. I circled it and we tried to fire the Verey pistol but none of the three cartridges went off.

I started to panic then. We climbed and I told the all crew except the navigator to bale out and they did. Then the navigator came and sat next to me. We flew east of Wagga as we didn't want the plane to come down anywhere near habitation. After a little while the motors cut out so the navigator got out and I followed.

I came down safely in a field. I just missed some trees and then hit the ground. The plane also landed in a field—more or less in one piece. We probably would have been all right if we'd all stayed in it. The tanks were absolutely dry—no petrol at all.

The whole unit moved up to Evans Head. We flew Ansons there and the work was the same routine exercises. There was a tendency for pilots to become browned off. At times they were tempted to let off a bit of steam.

One of the exercises took one pilot very near where he came from. He lived on a farm and decided he would shoot up his own home. He dived down with everyone on board, came very low and was circling when a wingtip hit the ground and he crashed. Everyone on board was killed.

He was unrecognisable when his family went out to the plane. They rang up to report a plane had crashed on their property. In the course of the conversation they said everyone was killed and did the station know who was the pilot. The station people said 'Yes,' and told them the name. The family asked if anyone else could have taken the flight. But the answer was 'No.' Their worst fears were confirmed.

Evans Head was a delightful place to be stationed because we could go down to the beach for a swim every day after we had finished. But there were crashes all the time.

PAPUA, EASTERN NEW GUINEA [MANDATED TERRITORY]

Early Stages at Port Moresby, 1942

Japan entered the war in December 1941 and Australia's defence position changed immediately. Instead of confronting a distant enemy, Australia found herself with an enemy, not only on her doorstep, but actually threatening invasion.

General MacArthur came to Australia from the Philippines and took over as Supreme Commander. Papua/New Guinea, our closest neighbour, was under threat of attack.

The position in New Guinea was critical in 1942. Not only was the decision to reinforce it taken too late, but, even when taken, the wrong forces were sent. A brigade of half-trained militia was despatched while the highly trained 7th Australian Division was left in South Queensland.

The administrative system for the supply of essential equipment was hopelessly inefficient and conditions for personnel were primitive.

Fred Cassidy has a very pertinent comment:
Living conditions weren't good—bully beef and biscuits and mosquitoes. We had two good RAAF cooks—two friendly boys. They were fairies but they kept a good kitchen. We didn't have separate Officers' and Sergeants' Messes for aircrew, but just one Aircrew Mess. It was felt that if crews flew together, they should also eat and live together.

Bruce Edenborough comments:
Things were very tense—it was the time of the Battle of the Coral Sea and one of our jobs was to ferry American soldiers to Port Moresby. We used to take thirty at a time.

The Americans had laid down metal strips and filled the place with fighter planes and all manner of other aircraft which we could only ogle at. Our resources were so poor that just the sight of all this made us all realise how good it was to have good and powerful friends.

Port Moresby, 1942.

The Australian effort seemed very secondary to the mighty American effort and all the magnificent planes and equipment.

Fred Cassidy tells of a prang on take-off at Port Moresby:
We were taking off from the end of Ward's strip, at the end of 1942, and were a long time waiting in line to get airborne. We took it a bit too quickly. Beaus were a bit difficult to get off the ground, and were inclined to loop if you didn't get your tail up early and get some rudder control.

What with the heat and being in a bit of a hurry, we spent a long time going sideways and cleaned up a few planes and a bulldozer on the side of the strip. We wrecked our plane but, fortunately, didn't get hurt.

Peter Fisken had a prang landing at Moresby:
We'd been attacking at Buna and had copped a lot of small-arms fire underneath which shot away the hydraulics. We couldn't get the undercarriage down but landed anyway. It was amazing how little damage was done—virtually only the props.

Port Moresby was the hopping-off point for most ops. The Beaufighter was very speedy at sea level but wallowed like a duck at high level. It was designed for low-level work although I understand it was used at high levels in the UK.

All our targets were over the other side of the Owen Stanleys so we had to stagger up through the passes and over the top.

The Air Defence of Port Moresby, 1942

Bob Crawford, of 75 Squadron, paints the picture:
At this time we didn't have many aircraft. Our air defences were very meagre but 75 Squadron shot down a bomber the day after their arrival. The next day they went over to Lae and damaged and destroyed lots of bombers.

One bloke, Wilbur Wackett, was shot down and walked back over the Owen Stanleys. It took him about three weeks. Johnny Piper got so excited he hit a bomber with his propeller and pushed the gun out the back of the wing. I think they lost two people on that raid.

Beaufighter.

We had to keep top cover over the aerodrome. When the other aircraft were on the ground we stayed up—there were always two up and two on stand-by. We had some interceptions and false interceptions, but not many because the army in the Owen Stanleys could see the Zeros take off from Lae. That meant the bombers were on the way. We could intercept them and do what we could.

John F. Jackson, a very good CO, was shot down over Lae while on a recce and had to walk back over the Owen Stanleys. He was picked up by American A-24s (Dauntless) from a little strip near Wau.

I was shot down over Moresby. I was No. 2 to Les Jackson flying top cover over the aerodrome. We got the alert that there was a raid coming in and our other aircraft took off and came up.

We ran out of fuel and landed at a little American strip right on the water instead of at Jackson Strip. We landed, refuelled and took off—we weren't on the ground more than thirty minutes. As we did a left turn over Moresby a B-26 (Marauder) went by with six Zeros on its tail. We attacked them although we were still climbing away with our wheels coming up. I don't know what happened to the B-26 but the Zeros forced me down into the harbour. Bullets were going everywhere round me and I finished up in the water.

The blokes who got me stuck around and circled overhead as I went in. When I came up I still had my parachute on. I thought they were going to strafe me but they attacked some Catalinas instead.

I had no dinghy but six inches of the rudder stuck out of the water and I sat on the tailplane, hiding behind it. A fuel barge which had been dispersed during the raid picked me up and took me back to the wharf.

Two others were shot down in that action. One crashed in the kunai grass west of Moresby and the other was killed.

I was hospitalised straight away but was in one day and out flying again the next. So was the other bloke.

Coral Sea

On 4 May 1942 a Japanese invasion force of eleven troop transports left Rabaul to attack Port Moresby. They were protected by, in all, three aircraft carriers—two large and one small—six cruisers and six destroyers.

Aware of the Japanese intentions—the Japanese naval code had been broken—the Allies were able to bring together an opposing force of two carriers, eight cruisers and eleven destroyers.

Jim Cowan claims to have been the first to sight this fleet, or at least one of the first. He gives his version:

There were only two squadrons available—Nos 11 and 20—and they didn't have enough aircraft to operate individually. So we flew in one another's aircraft with crews made up from members of both squadrons.

The Allies were aware that ships were moving as they had cracked the Japanese code. But they didn't know where. We had three Cats operating in the area that day and they had all been warned that they should get an enemy sighting.

We were out in the middle of the Coral Sea, about halfway between Milne Bay and Buka, which is 500 miles from Moresby, when we sighted the Jap ships coming up over the horizon. They were easy to see because they had great big forecastles and masts like a fifteen-storey building. I don't know whether we were the first.

We were down low because that is the best technique for doing a sea sweep. At 100 feet the horizon is only ten miles away so it is easy to see a ship coming over it and as an aircraft is much smaller than a ship you can see them much more easily than they can see you. Also it's more likely that the lookouts will be looking up high for aircraft. If you are high you can see a lot further but you can't see as clearly.

Our orders were to just report any sighting and get out—not to be gallant.

We continued with our search and came upon another group of ships—a couple of cruisers and two or three destroyers.

Then we found a solitary ship. It was very early in the war and none of us had done any ship recognition. We saw the ship and thought, 'You beaut. This will do us, we'll have a go.'

I was looking through Jane's Fighting Ships and suddenly recognised it as a float-plane carrier. We suddenly lost our aggressive spirit, turned round and got out.

Sometime after midday we were heading back home and having a bit of lunch after a busy morning. We were flying through low cloud of the Inter-Tropic Front. We suddenly came out in the clear and here, lo and behold, was a Kawanishi flying boat.

It must have given him a hell of a shock too. The skipper hit action stations and everyone rushed to the guns. But by that time he'd passed us and we'd passed him.

The Battle of the Coral Sea was not a naval battle. It was a sea/air battle. The fleets never came within shooting distance of one another. All the fighting was done by aircraft. One of the Cats got caught up in the middle of all the shuttling of aircraft backwards and forwards and was shot down.

Jim Cowan is right. The Battle of the Coral Sea was a sea/air battle. The two fleets did not encounter each other. Air attacks were the key factor.

Japanese carrier based aircraft attacked and sank the aircraft carrier USS Lexington *and damaged the USS* Yorktown. *The destroyer USS* Sims *and the tanker* Neosho *were also destroyed.*

The small Japanese carrier Shoho *was sunk and the large carrier* Shokaku *was damaged.*

Honours were roughly even, with the Japanese perhaps having a slight edge. However the Japanese invasion force had to withdraw to Truk and the attack on Port Moresby was called off. In the overall picture of the war in the Pacific, this was a most significant result.

There seems little doubt that Port Moresby would have fallen if the invasion force had been landed. There were few troops to defend it and the air defences were at a low ebb, as Bob Crawford points out:

Things didn't look too good at Port Moresby so they decided to evacuate us (75 Squadron) back to the mainland. We only had three aircraft left anyway.

All the pilots met in the mess that night and Les Jackson, the CO, said he was staying there and we could draw straws to see who was going back. Sufficient ground crew would remain to service the planes.

Those of us who were going packed up that night and left. Our escort was two corvettes to take us back to the mainland. They only went fifty miles and then left us. We got to Townsville in about four days.

The three who stayed, Les Jackson, Peter Marshall and Michael Butler, had their aircraft on the little strip beside the water. Then the Zeros came over the hill and strafed them. One was beyond repair. Michael Butler's could be repaired and the third was OK. The CO took the latter back to the mainland.

Michael Butler stayed on to get his plane repaired. He flew it back as far as Coen where he blew a tyre and left it. Peter Marshall hitched a ride to the mainland.

The squadron doctor (Dr Butcher) stayed behind. When the Coral Sea battle was over he did a quick air gunner's course and hitched a ride to Cairns in the back of an A-24.

We reformed and flew down to Bankstown in three or four days. There we picked up new aircraft and flew them back to Townsville.

The Battle of Milne Bay

Milne Bay is located at the south-east tip of New Guinea. It offered the Japanese a harbour for a thrust at Port Moresby. MacArthur wanted to prevent this and, in June 1942, ordered that an airstrip be built there.

There is little flat land there and much of what there is was overgrown plantations and broken by swamps intersected by small creeks. There was thick jungle near the coast and the roads were primitive.

The airstrip was built by the Americans and the 24th Field Company of the Australian Engineers. It was finished in July. The site for a second airstrip was selected at Waigani and a third was planned. By late August there were 9000 Allied forces in the area.

Jim Cowan flew in some senior officers prior to the Japanese attack and was struck by the lack of amenities for the troops:

When we got there morale was low. They didn't have writing paper or playing cards. They had no yeast to make bread, no toothpaste—none of the things you need. A fellow came up to me and said, 'Here's a couple of quid. If you ever come back here will you bring us back some books.'

I said, 'I don't want money—I might never come back.' 'Well, put it in the Comforts Fund—it's no good to me.' he said.

In fact we did get back, about a week later, and we had everything on that plane. We went to the Comforts Fund and said, 'We've got this group at Milne Bay (it was called Fall River then), give us what you can.' They loaded us with comforts—we even managed some cases of beer which took a bit getting through the blisters.

We landed on the water and there was no wharf evident so we went ashore in the dinghy. There were boats on shore to take the stuff in. They'd knocked the tops off some of the beer before they hit the shore.

The Japanese landed on 25/26 August. Bob Crawford was there with 75 Squadron (Kittyhawks):

It was called Fall River in those days. The strip was gouged out of a coconut plantation. It was very wide bordered by trees and ran roughly east to west into the bay. It was dirt with metal matting—a good strip but treacherous in wet weather.

We weren't there long before we started patrols and the Japs arrived very soon after. On that night four of us took off, flew west then north and were going up towards Kokoda when we were told that barges were coming into Milne Bay.

The weather was bad but, luckily, we caught them just coming to land on an island. We destroyed the lot (about ten) and kept strafing until we ran out of ammunition.

The weather was very bad at Milne Bay. On average there was cloud at 1000 feet and heavy showers of rain. You just had to go round them. We'd only go through cloud when over water as it could be fatal if over the mountains.

We came back in the afternoon and were told that a Jap convoy was coming in. It had been spotted by Beauforts who had been out to attack it with torpedoes. 75 and 76 Squadrons took off to attack it. We flew under the cloud and sank a gunboat.

As we came back one pilot came in the wrong way, landed in the coconut trees and was killed.

The convoy landed troops that night and all hell broke out next morning.

Les Jackson took off with four planes at first light and attacked the invasion forces. I was on the target continuously—take off, strafe, land, take off, strafe . . . Where possible we attacked barges but we also attacked troops coming down the road, fringed as it was with plantations. We bombed as well as strafed and Peter Turnbull, CO of 67 Squadron, was killed on a bombing run.

We did this continuously for three or four days. If we weren't flying we were belting up ammunition as we were using so much. The ground staff were flat-out refuelling. The conditions were appalling, mud everywhere and rain pouring down.

The Japanese were supported by a cruiser. It would buzz off each morning but come back at night and shell us. The shelling went on all night. With the low cloud you'd hear the 'Whoomph' and see the flash reflected as the gun fired. We'd see the flash, hear the bang and count twelve seconds. We reckoned that if, after this, we could hear the 'whirr' of a shell going through the air we were all right. If we didn't hear the 'whirr' we were in for trouble.

There were no shelters because we'd had so much rain that all the slit trenches were full.

We ate down at the strip in the daytime and at night the kitchen staff would have bully beef ready and a smiling conversation. Then, relaxed and showered, we'd go to bed.

It was very dicey. Things were getting critical. The Japs were making progress. The army blokes did a wonderful job—many of them were only part-trained or were new to jungle fighting.

We had to evacuate the two squadrons out of Milne Bay because there were fears that the Japs would overrun the strip. Our forces were building another strip at Gili Gili.

The squadron flew out but Bill Cowie and I had to wait as our aircraft were being repaired. We flew out and I got back to Moresby about 5.30 p.m. It quickly got dark and Bill came into the circuit, went down-wind, cross-wind and turned into the hill and was killed.

We were at Moresby for two days. In that time the army turned the Japs back from the strip that was under construction. I was flown back in a Lockheed Hudson as I was a surplus pilot. The ground staff had been left at Milne Bay—only the planes were brought out as they were essential to survival—so we went back to our strip.

Bombs were brought in by ship. We had plenty of ammunition but we used so much when the Japs landed we had to have more flown in. An American Liberator landed on the strip and unloaded ammunition but went unserviceable and the Zeros came in and set it alight.

While the invasion was on the Anshun used to come in at night and buzz off before the cruiser came in. It was slow getting out one night and the cruiser came in and sank it at the jetty.

We had an A-24 (Douglas Dauntless dive bomber) come in at 8 o'clock one night. It had been bombing the Kokoda Trail, got separated and flown to Milne Bay. They were attacked by Zeros and the air gunner (a Red Indian) was killed and the pilot hit. The plane was riddled. Luckily we had an army hospital there.

I contracted malaria in Milne Bay but it didn't catch up with me until later.

Finally, on 6 September, the Japanese forces withdrew. They had met their first defeat.

Thereafter the Milne Bay strip was used as a base for 100 Squadron Beauforts.

Len Parsons tells of some of the operations from there and of the difficulties of life at Milne Bay:
My first operational posting was to 100 Squadron in October 1942. I joined it at Bohle River, just at the back of Townsville. We did our little bit of training at Townsville Harbour and then flew to Milne Bay. Training was only a matter of a few weeks.

Milne Bay was a level strip but was very churned up at that time. I remember we got there one afternoon. The Japanese must have known we were coming because they arrived half an hour later.

It was a pretty good initiation. They concentrated on the strip and the bombs went straight through it. Luckily it was wire matting and could be repaired quite quickly.

The slit trenches hadn't been used for some time and were half full of water. The ground was so soft that bombs buried themselves in the mud and left little sign of where they landed. We saw them land about twenty feet from us.

The next day the CO took twelve of us out in formation. I believe there was supposed to be a Jap convoy landing supplies at Buna. We turned back when about three-quarters of the way there. I don't know the reason but I believe it was counted as a strike.

Then we did a lot of reconnaissance work looking for Japanese submarines and ships.

We could carry six bombs or a torpedo. We carried bombs on reconnaissance work and jettisoned them if attacked. One of our blokes was attacked by four Zeros and that was the first time an Australian aircraft had been able to fight off a concerted attack by four enemy fighters. In those days we had an extra gun at the front and two .5s in the rear as well.

The Beaufort was really a fantastically manoeuvrable aeroplane. Unfortunately we were going through a stage when we lost a lot of them. There was a problem with the elevators. The controls were locking in the down position.

It happened mainly only on training aircraft—you'd be flying along normally and it would suddenly dive into the sea. On operations we always had new aircraft.

We mainly did reconnaissance—only five attacks in the period.

Battles of the Kokoda Track
July – November 1942

The Japanese landed troops at Gona and Buna, on the northern coast of New Guinea, on the night of 21 July 1942 with the aim of crossing the Owen Stanley mountains and taking Port Moresby.

Lieutenant Colonel Owen of 39th Battalion took command of Maroubra Force at Kokoda on 21 July. His task was to prevent the Japanese reaching Kokoda but, if this was not possible, he was to hold Kokoda.

Kokoda was situated in a valley 1200 feet above sea level and was windswept and rainswept. The hills were precipitous and densely forested, there were no roads except for the twenty-five miles from Port Moresby to Sogesi. Transport was by porters along the steep and winding track although horses and mules were used in the early stages (the Port Moresby end) of the track.

The Australians retreated, yielding Kokoda, Myola and Iorabaiwa before the Japanese were halted at Imita Ridge overlooking Port Moresby.

30 Squadron RAAF (Beaufighters), within the limits of the planes available, supported the army in the retreat. Fred Cassidy took part:

The Japs were only seventeen miles from Moresby at this time. The Liaison Officer would come in and say, 'We want you to strafe on a line 500 yards north-east of a bend in the Tumesi River. You can't miss it. There's a little clearing, there's a big tree and there's three huts and it's just on the bend in the river.' He would give us a bit of a latitude and longitude and say, 'That's where it is.'

You'd go up there, and, from the air, you could see forty-five bends, fifty little clearings, a hundred trees that high—which one do you pick? To that extent it was funny, yet it was serious, because quite often, when the army fellows came back from that area, they would want to see who was over the area on a certain day. They would have with them the shells from the cannons and machine-guns—our empty shells which had fallen on top of them.

It got to the stage where the points had to be identified a bit better. They started using smoke mortars to give us a better idea. The theory was that we'd be able to see the smoke coming through the trees. However, there was always smoke coming out of the jungle! So smoke mortars weren't much help.

The Japanese began pulling back in late September. First to Iorabaiwa and then the retreat quickened. The Australians pushed the invaders back to Buna and Gona which were to become the sites of bloody campaigns.

Beaufighter with 30 Squadron members, from left, R.Wilson (navigator) T.Butterfield (pilot) both killed in action, October 1942, and Fitter I.Gazzard.

Captured Japanese landing barge at Buna.

Battles of Buna, Gona & Sanananda November 1942—January 1943

Australian forces were committed to a campaign to capture Buna, Gona and Sanananda. MacArthur made the false assumption that the Japanese were ready to evacuate and could be easily defeated. This was not the case. It was the Japanese intention to hold Buna until the outcome of the Guadalcanal campaign was determined.

Australian troops started assembling for the attack in October 1942. Gona was attacked on 16 November 1942 by 25th Brigade and Buna and Sanananda on the same day by the 16th Brigade.

Beaufighters flew strafing sorties in which Peter Fisken took part:

Buna was our first target. We lost our first Beaufighter to fire from Ack-Ack Charlie sited at the end of the strip. Ack-Ack Charlie was there all the time in spite of many attacks. He had something like a Bofors gun.

We were strafing aircraft, troops and barges round the strips. We never went in and blazed away at the jungle, we attacked down the strip and round the perimeter. We could see barges in the mangroves even though they were camouflaged.

We had one engine shot out in one attack on Buna. There was no way we could get back over the range on one motor so we went down the coast to Milne Bay and followed the coast round from there.

Jim Cowan tells of his experience in attacking Gona and Buna in a Catalina:

We attacked the area at night. My brother was shot down there in a Hudson and didn't survive.

Two nights after his death (I didn't know of it at the time) I was going up and down strafing the beach. We were at about fifty to one hundred feet. We had a fair bit of firepower—one gun in the rear, two in the blisters and one in front. We could see the shoreline because of the phosphorescence.

Fred Cassidy remembers this phase clearly:

We did lose a few fellows over Sanananda. We were strafing barges used for replenishing the stores up and down the coast, from Lae, Salamaua and round the bend to Buna, and Sanananda and Gona. The Japs held most of the north east coast of New Guinea.

Battle of the Bismarck Sea

In March 1943 the Japanese aimed to reinforce their forces in Lae with 7000 troops from Rabaul. This involved transporting them across the Bismarck Sea.

The force of eight transports and eight destroyers was sighted by aircraft on 2 March. The sighting led to an attack using all the aircraft which could be drawn from New Guinea and the Australian mainland staging through New Guinea. It went on for days and resulted in the loss of twelve vessels (only four destroyers survived) and 3000 Japanese.

Fred Cassidy, Peter Fisken and Les Parsons took part in the operation.

Peter Fisken flew a Beaufighter:

It was a well-planned attack under American control. Beaufighters were to go in at sea level to eliminate the anti-aircraft fire, then the Bostons would skip bomb and Liberators and Lightnings would come in at a higher level.

We came in so low the guns could not depress to get us and strafed the ships. Then the mediums dropped their bombs. One hit a destroyer just as we reached it and blew great holes in the bottom of the fuselage. We shouldn't have been there but we got enthusiastic and went back for another go.

Les Parsons was in a Beaufort:

The Japanese had been able to get pretty close before they were picked up—they had been hidden by bad weather. We had been on alert for a couple of days—not allowed back to camp from the strip. I had slept for two nights on the camouflage net over an anti-aircraft gun. Then we were allowed back to camp but at one in the morning the ships were picked up.

I was duty officer and, unexpectedly, the CO said, 'You're on this trip too.' So I had to go back to the tents and get my crew out of bed. The briefing was going on while I was doing that so I missed all but the tail end.

I missed the flares marking the convoy as I ran into a tropical storm. It was all right for planes coming up from Port Moresby but only four of us got through from Milne Bay. Two dropped their torpedoes—without result. At dawn I came down to sea level and found the altimeter was out by 200 feet. That error would have affected all altimeters and all the torpedoes would have hit the water at an angle of 60 degrees and gone straight down. No wonder they didn't see any results.

There had been a trial run of the bombing and strafing attack using a wreck at Moresby and it had been a mess. The attack that now went in was properly coordinated—Beaufighters in low first, then the mediums and the high-level bombers two minutes after them.

After our first try, we Beauforts went out the next day to attack any barges that were about. There were twenty or thirty and they were firing back with rifles.

The end result was that the Japanese reckoned they lost 3 000 men, all the transports were sunk or severely damaged and the majority of the destroyers were sunk. All in all a very successful day.

Another contributor, who wishes to remain anonymous, adds:
The squadron was sent out the following morning to strafe all the lifeboats and liferafts. It was not a thing I liked doing. There were few barges—they were mainly used up and down the coast.

None of the troops got ashore. Four 20 mm cannon and six machine-guns make a hell of an impact.

As far as I can remember we didn't lose any aircraft over the two days.

Fred Cassidy:
We attacked invasion barges during the Battle of the Bismarck Sea. There were twenty to forty barges full of troops. War is war and we had to help our army. They were armed so we strafed and bombed them.

Japanese ships ablaze during the Battle of the Bismark Sea.

Lae as a target

Lae is situated on the Huon Peninsula on the northern coast of New Guinea. It was invaded by the Japanese on 8 March 1942 at the same time as Salamaua was attacked. The town had been evacuated as it could not be defended.

The area appealed to the Japanese as they could base aircraft there for attacks on Port Moresby. It was always a hotly defended target for RAAF aircraft.

During this early phase of operations in New Guinea, Fred Cassidy found out just how vulnerable the Beaufighter was:
On my first tour we had no rear gun in the Beaufighter. We'd try to create a diversion by getting out the Aldis light [a signalling lamp] and flashing it at a potential attacker. Then the squadron devised another trick. We'd save up all our bits of coloured paper, tear them up and put them in a bag.

When I went into Lae, for instance, and was coming out, I threw these out through the flare chute to create some sort of diversion. I hoped to take the eye of the Zero pilot. Flashing the Aldis with one hand and throwing out paper with the other—it was so naive.

During this period, Cassidy operated on convoy attacks:
Beaufighters were given the role of knocking out the defences of the ships carrying men and equipment to Lae, so that the bombers, the Mitchells and Bostons, could get in close. We were to go in first and shoot up the bridge. These were the tactics that Garing [Officer Commanding 9 Operations Group 1942-1943] had devised. We weren't to come in sideways, but come in, head on, over the bridge. Garing said, from his experience, if you put the captain out, you put out the organisation of the ship, and you then had a better chance of wiping out the ship. So, that's exactly what we did.

We were strafing, down low, with cannon and machine-guns, as bombs came from on top from Fortresses and Liberators. They were lucky to hit any ships from that height. I think that most damage resulted from the skip bombing by the Mitchells and Bostons. We had to watch out for the bombs from above and from the skip bombers behind us.

We only had the opportunity to do about three runs in an attack. In and out again in an afternoon, hitting anything that was handy.

Once, the weather built up and we could only get about three-quarters of the way over the ranges before we had to turn back.

Another time, we left ships burning but didn't see any sunk. We couldn't hang round. We'd done our job and disorganised the convoy. We'd made them turn and leave themselves open for the bombing.

The Japanese ship *Moyuko Maku* damaged beyond repair after air attacks on Lae Harbour.

Fred Cassidy flew on missions out of Moresby:
The diciest airport to attack was Lae. The weather was always bad, it was heavily defended and the Japs always knew you were coming - we had to fly there from Port Moresby. It had fighter cover and had the heaviest ack-ack defences in the South Pacific.

The end of the strip came straight down to the sea and, to attack, we had to come in from the land side. The tactics were to have eight to ten aircraft on the target. Half would come in from one side, and half from the other side of the strip, just missing each other either way. The Japs had pockets of machine-guns, about forty feet apart, which pointed straight up into the air. They were located along the whole length of the runway, and you had to get through that to strafe the planes on the side of the strip. We got quite a few that way.

You always got hit at Lae. There was always a little hole in your plane when you came back. You'd be about 300 feet when you commenced your run, but as you came along you were only about forty feet up. When you got to the end of the strip you had to deflect left or right to avoid flying into the sea.

If you came down, you not only had to escape from the aircraft, you had to escape the Japanese, and at this stage they were only a few miles from Port Moresby. So you had to escape capture and the fate of Newton and his mates—i.e. execution.

On top of that, you had the rugged terrain of the jungle. You just couldn't find a flat spot to put down unless you were lucky, and then, once in, you had to get out. It was a very difficult part of the world to operate in.

Cassidy experienced a very determined fighter attack:
When I was with 30 Squadron Beaufighters, we made an attack on Lae. We were pretty fast on the deck and fairly manoeuvrable. We were only hit once, and then we were jumped by three Zeros.

When this sort of thing happens, you haven't time to go left or right—you can only go straight ahead. Here we had one on one side, one on the other, and one down the back. We were like a sitting duck. I was in the back, flashing my Aldis and throwing out my lolly papers. A bullet went through my wireless unit, went in one side and out the other.

These blokes were very clever. They never wasted their ammo. They had two goes on either side and we were skipping underneath them. We could see the bullets going into the water behind us and over the top. We were going at about 330 miles per hour on the deck and, as soon as they ran out of the impetus of their dive, they formated on us. Two of them, one on either side, stood out about 100 yards from us. They couldn't go any faster than we could, so they just waggled their wings and away they went.

They were navy pilots, in those days, and were the pick of the navy. I had always thought the Japs were bits of monkeys, but these pilots were real professionals. That's why I never liked going to Lae. It was the worst target.

Jim Cowan was flying a Catalina over Lae on night flights to harass the defenders:
We used to carry eight 250-pound bombs inboard, bottles (to distract the searchlights) and bundles of incendiaries (which we dropped by hand). We stayed over the target for four or five hours.

The ack-ack could only reach 8000 feet so we flew at 10 000 feet.

Our Intelligence Officers told us that the war casualties from malaria were higher than those from our attacks. So if we could keep the Japs in their trenches all night we could disable the whole bloody army. We therefore bombed at night.

The mere fact that there was an aeroplane up there which dropped something every now and again kept them in the trenches and that was why we had all the ancillary gear to drop. Otherwise they would just have gone back to bed.

Fighters could take off at night with a lead light that wasn't visible from the air. But they couldn't land without a flare path. So they wouldn't take off until they were sure they had enough endurance to wait for daylight. We used to stay until just before the possible take off time and then go home.

Peter Fisken has this to say about his flights to Lae:
Lae was an army base and the airfield was most strongly defended and heavily used. There were strong anti-aircraft defences and fighters could be a problem if they got off the ground.

We had an advantage in the quietness of the Beaufighter (it was known as the Whispering Death). We could come over the range fast, low and quiet and be there before they could get off.

That tactic failed once when we had Yanks as top cover. There was supposed to be radio silence but they chattered incessantly. About four Zeros were up before we got there. Two of them dived at us and the pilot tried everything to shake them off.

We had a useless gas-operated gun in the cupola but, luckily, it had been loaded with tracer bullets this time. As soon as a burst was fired at them they sheered off. Apparently they were scared of the tracers. We only got a couple of holes.

Len Parsons visited Lae while at Milne Bay with 100 Squadron:
We had eighteen planes in a squadron and in the time I was with 100 Squadron (nine months) we lost 121 blokes. The flying conditions were bad—the worst in the world.

In the wet season the clouds came right down to the ground. The rain was heavy and waterspouts went from sea level up to 2000 feet—you couldn't get through them. The cu-nim went from 2000 to 20 000 or 30 000 feet. We had no navigation aids.

I went to Lae one night in dreadful weather. I had the only radar set on the squadron for test purposes and decided to use it to get through the cloud. We picked up the coast round about Buna and followed it on radar.

I got to the neighbourhood of our strip but there was no break in the cloud. After four and a half hours of instrument flying I was thinking very seriously of putting down on an emergency strip. I was coming down when I saw a gap at 6000 feet and finally got through at 2 000 feet and landed at East Milne Bay strip.

There wasn't a cloud in the sky south of Milne Bay. It had been like that all night.

Although I had been wandering up and down the coast all night I was the first to land. The others had set course for Lae but had finished up at the Trobriands—the wind was 150 knots from the south and put them 30 degrees off course.

Tsili Tsili

An airstrip was hacked out of six-foot high kunai grass at Tsili Tsili in early June 1943. The drome went into service on 26 July 1943 when the first fighters landed there.

In early September 1943 24 Squadron was rushed up to the new aerodrome. Ron Barker was a member of the squadron:
On the last day of August 1943, a request came from MacArthur for a dive-bomber squadron to be used in New Guinea. We were partially equipped, at the time, with four Vultee Vengeance aircraft.

The message was received on Friday and on Monday twelve aircraft took off from Bankstown to fly north. In those two or three days, we had been equipped with the remainder of the aircraft of our squadron.

The squadron moved up to New Guinea as a detachment of fifteen aircraft. We flew from Bankstown to Amberley, where we staged in with an American unit. Then to Cooktown, to Horn Island, and then to Moresby.

Our navigation was by visual sighting all the way. It was August, and the weather was completely clear. It was pretty much a matter of just following the coast.

The first operation was on 7 September, and others followed soon after. Ron Barker comments:
Our trips were, normally, about one hour forty-five to two hours thirty, but our first trip was not typical. We were called on to fly into the Markham Valley and bomb the bridges on the road which led down to Lae. The road was only a dirt road, winding up, through the valley, but it was used by the Japs to bring reinforcements down to Lae.

At that time, the Americans had planned to use the Australians to land at Lae, and they wanted us to bomb the bridges to stop reinforcements coming down.

We went out and followed the valley up to 10 000 feet. We located the road, but had difficulty in locating the targets, because the route went through a jungle area.

Although there were four airstrips, the weather closed in and we couldn't find any of them. So we kept the bombs on board and, as instructed, started to look for an aerodrome that the Allies had behind the Japanese lines. It was in a landlocked valley in the mountains and was hard to find. We were flying in among mountains and there were valleys wherever you looked.

We had flown up and down, looking for the valley when the CO came up on the air and said it was every man for himself. If you couldn't find the aerodrome, put down in the valley. That would have been easy enough, it was grassy and flat, although it would have meant a wheels-up landing. As it was in enemy territory, we weren't keen on it.

My pilot said, 'I can't find the strip. What do you think we should do?' I was a pilot officer, and he was a flight sergeant, but he had to make the decision, so I suggested he should gain altitude and see if he could see over the ranges. He did this and, as he gained height, he could see another of our aircraft going down into a valley. We followed it and, sure enough, there was the land-locked valley and the airstrip—and a heavily armed American base.

Our twelve planes landed on the strip—most with the red light showing we were almost out of fuel—very dicey. The total time for the trip was four hours and twenty minutes—right on the endurance limit for our aircraft.

Few squadrons were equipped with dive-bombers, so a note on techniques from Ron Barker is most instructive:
Our main enemy in New Guinea was the weather. At 9 a.m. we could be at the strip, waiting to take off in bright sunshine. Within an hour, the raid could be cancelled because the weather had built up over the target area or between us and the target area. It happened several times. We were in this area of New Guinea during the months of September to March, which is the wet season.

We weren't equipped with radar, but had radio and IFF [A device to identify friendly or enemy aircraft]. Our armament was twin .303s on a swinging mounting in the rear, and four .303's, firing forward, fixed in the wings - two in each wing. The bomb load was two 500-pound bombs in the bomb bay and one 250-pound bomb under each wing.

Dive-bombing was a particular strategy that had been developed over many years. It commenced in America in the 1920s—I believe Major Mitchell developed it. The Germans perfected the art and the US Navy, but not the army, continued it.

We had learned one method of dive-bombing in

training. We'd fly up to the target, the leader would come up and over and go into a dive. We would start at anything from 10 000 to 14 000 feet.

However, in practising this at Bankstown, a couple of planes collided. So we had a couple of American colonels, who had done some dive-bombing at Guadalcanal, come over and give us some instruction in their technique.

After this, we'd approach the target at about 4000 feet. The leading aircraft would slip away underneath and then go down in a dive. Then, gradually, one by one, the other aircraft would follow. Each plane would turn over on its back as it went down.

Being capable of a 90 degree dive, each pilot would manoeuvre the aircraft on the way down until it was aimed at the target. He would use the gunsight. We'd start to pull out at 3000 feet and eventually squash out at 700 feet. We would put the airbrakes out as we were going into the dive. These would hold us at a steady 300 knots all the way down. If the airbrakes failed we had to go in without them. But then, it would be more difficult to pull out of the dive. I've been in the plane at Bankstown when we tested it without airbrakes—we reached a speed of well over 400 knots before we pulled out. [The airspeed indicator was calibrated in knots as the aircraft was intended for naval use.]

While you were going down it was just like a normal flight, but when you pulled out, all the blood would drain from your head to your feet. You didn't get to the black out stage, but you'd be very close to it. Some did black out on occasions, and that was dangerous. The force of gravity was enormous, your whole body seemed to be gripped by the bottom of the aircraft. You couldn't move your legs or hands.

It was acknowledged that it would be very difficult to get out if you were hit on the way down. The plane would have to come out of the dive so you could slide out over the side. We had seat parachutes. On top of the parachute was an inflatable dinghy with a compressed-air bottle which bit into your bones and made a very uncomfortable seat. But that was something you had to put up with.

It was a very heavy aircraft. It weighed about eight tons. But it was a very easy plane to fly. I found, at times when I had the controls, that they were very light. Pulling out of a dive required some strength on the part of the pilot, but I never discussed it with them to find out how they felt about it.

Ron Barker emphasises the problems of navigation in the New Guinea operations:
Our navigation was by visual contact. I only used dead

reckoning when I was flying over the ocean. Usually, there were a number of aircraft flying in formation and the lead aircraft would always have the lead navigator.

I remember being in the lead aircraft with the CO on one operation. Although we had to go out to sea to attack an island off the coast, it was a trip where we could make good sightings. I don't remember doing any dead reckoning navigation on that trip.

It was hard to get a pinpoint over the jungle. You would set your compass course as you set off and follow, as best as you could, on your map. It was very difficult to find features on your map because the valleys and mountains all looked alike. But there were some features that helped. For example, a mountain peak. From your flying time and airspeed you could find the distance travelled and identify a peak.

Even when flying in formation, it was the duty of the navigator to know his position at all times. We always did our own navigation—not to guide the aircraft, but in case we were attacked and got separated, we could find our own way home.

Kiriwina Island — Trobriand Group

Kiriwina Island and the Trobriands had been occupied by the Japanese and some of the survivors of the convoys meant for Milne Bay had landed there. The mission given to Jim Cowan and his crew was a risky one:
While flying we got a message to proceed to such and such a position and attempt to rescue a crew which was down. It was near a small atoll in the Trobriands. We flew around the group two or three times and couldn't find anyone or see an aeroplane. We landed in a lagoon—near a village.

We couldn't communicate with the natives and decided to do one more sweep before we left.

As we swept round a fellow came out on the beach and waved at us. We landed on the sea outside the lagoon.

It transpired that the survivors hadn't realised we were a friendly aircraft and had not wanted to let people know where they were.

They had pranged off the island and must have been close in because they swam ashore. That was why there was no aircraft to be seen.

Extraordinarily, the fellow who came out in a canoe had been an office boy for one of my friends in Adelaide. I said, 'What he hell are you doing here?' He responded in the same fashion.

Because the American aircrew were, at that stage, inexperienced in local conditions it was general practice

to attach some RAAF aircrew to them. He was one of those attachments.

About three days later we went to pick up Bob Gurney who had gone down in that area—he was a good mate of mine. We found the aircraft but there were no survivors.

Air Attacks on Salamaua

Salamaua is situated on the Huon Gulf on the northern coast of New Guinea. It provided a good port and the Japanese landed without opposition on 8 March 1942.

The airfield was poor and bordered a large swamp but was a menace to the Beaufighters of 30 Squadron while the Japanese held it.

Salamaua was recaptured by Australian forces in September 1943.

Fred Cassidy:
I also remember Salamaua, which was a pretty big place for New Guinea. It was just north-east, between Gona and Lae. There were big naval guns on the headland. As we came past, these naval guns would open up. Every time they went off I'd say 'Duck!'—although they were a long way away. After I'd ducked about six times I woke up to myself.

Jim Cowan had a very satisfying attack on it in his Catalina:
I remember one night when we went to Salamaua. It was on a long isthmus, about 200 yards wide, and broadened to about 500 yards at the end.

I dropped a stick of bombs right along the peninsula where the town was. They went down one, two, three—spot on.

Finschhafen

As part of the general objective to 'seize the Lae-Salamaua-Finschhafen-Markham Valley area' an attack on Finschhafen was planned before the fall of Lae. By 16 September 1943 the Australians occupied Lae, Salamaua and the Markham Valley east of Nadzab. Planning for the attack on Finschhafen was accelerated.

It was decided that 20th Brigade should carry out an amphibious attack over Scarlet Beach. This would be supported by Vultee Vengeances and Bostons attacking targets in the Finschhafen area.

The Vultee Vengeances of 24 Squadron were used in these attacks. Ron Barker describes this:
We were bombing, mainly, on map positions which showed us where the enemy was located. We couldn't see clearly at 10 000 feet, but, when we identified the general area, we came down and could then see things.

Vultee Vengeance dive bomber.

The anti-aircraft fire was heavy in this area so we didn't stay round to see the results of the raid—we had only the results of our own limited observations. When the army went in, some three days later, they found that the place had been devastated by our Vultees.

On our way back from that raid, we flew halfway across the Huon Gulf and over Morobe, where there was a town and an Australian base. At the entrance to the harbour were two American destroyers and, unfortunately, the CO led us directly over the top of them at 8000 feet. I was facing the rear at the time, looking for enemy aircraft, when I felt a crump, crump, and the aircraft started to rock. I looked round the cockpit and saw little black lumps of soot floating around. I thought, 'There's no fire.' I looked between us and the next plane and saw a puff of black smoke. It was then I realised that we were under attack from the AA. I looked over the side, and there were the two destroyers firing directly at us. Flak was also coming from the land batteries—the Australian Army was firing at us as well!

They had us completely trapped, but the twelve aircraft immediately dived to the deck, right down on the water, safe from the flak. As we were flying south I could see the shrapnel falling in the water behind our planes.

Wewak

Wewak is situated on the north coast of New Guinea. It had a good anchorage and the hinterland was suitable for the development of airfields. It was the administrative centre for the Sepik district at the outbreak of war with the Japanese but was evacuated in April 1942 in the face of Japanese attacks.

In February 1943 the Japanese strength at Wewak was estimated at 9000 men. This grew when the Japanese XVIII Army fell back on the town and consolidated there in March 1944.

This was the position when Len Parsons attacked Wewak:

The squadron was based at Nadzab in the Markham Valley. We were attacking Wewak which was strongly defended but the situation improved when the Americans caught about 160 aircraft on the Wewak strip and made life much easier.

The round trip took about four and a half hours. We came in from the sea to avoid as much flak as possible but I remember one morning we came in at about eight in the morning at 8000 feet. We were hit almost at once. The plane alongside me disappeared. I don't know what happened to him but we dropped our bombs.

Our aircraft was not severely damaged. The flak had gone through the front near my feet, through the

Reconnaissance photo of Japanese anti-aircraft gun emplacements issued to pilots before a raid. Wewak 1944.

navigator's table and out through the roof. We were lucky. Another squadron was due to come over immediately after us so I hung around to see the results.

When I got back I found the crew which had been shot down had been picked up by a Catalina and was already back home. There was always a Catalina at the target area and it was a great help to know that, if you could get far enough out to sea, you would be picked up and brought home safely.

Wewak was recaptured in mid–May 1945 but pockets of Japanese resistance still held out to the south.

Les Parsons comments:

We attacked villages near Wewak. At first we found our targets with the help of the army who fired smoke shells. Later they used marker aircraft—mainly Wirraways.

They marked the target and we would bomb the markers when they gave the signal that the markers were in the right position.

Finding the target was a bit worrying even though we had an army liaison officer on the squadron. I've known people who have bombed their own troops about five miles away from the target. It was very difficult to pick one ridge from another and the Japs could be on one side of a ridge and the Australians on the other.

Our raids were mainly bombing—we used anti-personnel bombs.

Dobodura Airfield, 1945

When Buna and Gona fell in December 1942/January 1943 it became possible to build a large airfield at Dobodura. This brought Allied aircraft closer to their targets in Japanese-held territory to the north and north-west.

Bruce Edenborough operated in Beauforts from here in May 1945:

Our crew was posted to No 6 Squadron at Dobodura. We were ferried to Townsville and crossed over to Port Moresby in an American flying boat. We were then ferried to Dobodura.

There we were right in the middle of American operations. We messed with the Americans, enjoyed their good food and found we could exchange our old khaki dress for good quality shirts and shorts in the American store. Our gear seemed to wear out remarkably quickly and we were soon fitted out in the great American uniforms.

We found that the war had really moved on from Dobodura. All the damaged countryside and wrecks were there but the war had gone up further. We were in New Guinea but there was not much really going on.

We kept up training. I once ferried a group of Americans to find some forests and identify the trees in them. I couldn't see what that had to do with the war.

Then there was a lot of photographic work—mosaics of the country up behind Wewak which was still being bombed. The Australians were mopping up the Japanese in that area then.

We did convoy anti-submarine patrols. We once were wakened at 3 a.m. for a 4 a.m. take off to escort a convoy. We were given its location and flew up and down the coast with our depth charges under our wings for hours looking for it. Without ever sighting the convoy we landed at Madang at eight o'clock to learn that it had never left Milne Bay. This was just before the attack on Tarakan and Balikpapan and showed just how poor was the liaison between the air force and the navy.

We were rather put out when we were told we were being transferred to another unit. The heyday of our squadron was past.

Servicing Problems

Servicing aircraft in the tropics gave rise to many problems. Ron Barker tells of some of these:

We had some mechanical problems. One plane caught fire, for no apparent reason, over the sea and the pilot had to put it down in the water, about 100 yards from the beach. It sank to the bottom in fifteen feet of water. The crew got themselves out, came to the surface and swam ashore. They were rescued by the army and got back to base later.

Another pilot had engine problems over the target area—somewhere up in the mountains. He had difficulties in changing tanks. He landed in the Markham Valley, on the grass. The crew sat there for three or four days before a light aircraft came up, landed nearby and took them back to base.

On at least three occasions our fuel pump packed up and we had to abort the attack. There was a hand pump in each cockpit and we took turns to pump petrol and fly the aircraft all the way back. We were so busy, I couldn't look out for enemy aircraft or man the gun. We were very tired by the time—about an hour—we got home.

Ferry Flights

Planes had to be ferried up from Brisbane to maintain the strength of the squadron, as Ron Barker of 24 Squadron (Vengeance aircraft) observes:

We observed radio silence at all times. The only time I had to use the radio for an emergency was when we were ferrying a plane from Australia to New Guinea.

We were flying at a couple of thousand feet when a flame shot out of the engine, which then started to run rough. We were midway between Cooktown and Port Moresby. A lot of aircraft were lost in that area.

I immediately put out a Mayday call on the emergency frequency that had been given to us before we left. I also put out an SOS on the key. I hoped they would take a fix on us.

Fortunately, because I didn't get any response on the radio, the engine recovered. I was navigating over the ocean by D/R and happened to come out right on target.

There was no interference on the radio, but I did not receive any acknowledgment of my signal. When we arrived I checked in and reported that I had sent a Mayday message, but they said they had not heard me. It then transpired that I had been given the wrong frequencies at Townsville.

There was an Air Sea Rescue Service in operation, but it wasn't fully developed until later—this was in 1943. If you went down a Catalina or a P/T boat might pick you up—it did happen on several occasions. We were told that, if we came down near the coast, we should get into our dinghies, get to the coast as quickly as possible, and hope to be picked up.

We were halfway across to the Trobriand Islands once, when I saw a dinghy floating in the water. I had a good look as we passed over it, but there was no-one in it. There were several sharks nosing about.

Bob Crawford flew with his squadron (Kittyhawks) to Port Moresby:

I was posted from Darwin to 75 Squadron but nobody knew where it was. We were supposed to have been there fourteen days earlier but the signal got mislaid in the general confusion at the time. The new CO of my old squadron (Black Jack Walker) was upset to lose two experienced pilots, particularly as we had been training the new pilots. But within three hours of receiving the signal we were on a DC-3 and on the way out.

We got to Brisbane next morning and, after a year in Darwin, decided to blow through for three days. We could have stayed a month but we ran out of money. We caught a train to Townsville on the Wednesday and arrived on the Friday.

The squadron was flying Kittyhawks. We saw the CO on arrival and he said, 'We're leaving this afternoon to fly to Port Moresby.' I said, 'Well, we're sorry, but we only found out about this on Sunday.' He asked how long we had been in Darwin and we said, 'A year.' He said, 'Well you'll need a bit of leave.'

We left on leave. Michael Butler, who had accompanied me, lived in South Australia. He and I went to Sydney. We left on the Friday night and got to Sydney on Monday morning. At 2 p.m. the same day we got a telegram requesting us to report immediately to 76 Squadron at Archerfield.

We were on the plane that night. We got to 76. They had four or five Kittyhawks and, from what I could gather, they were going to put them to use as soon as they arrived in New Guinea. Their headquarters was located in a spare parts crate and they were getting pilots wherever they could. We were given priority and I think I got four to six hours on Kittyhawks.

In three days we were on our way north. We flew with our own 75 Squadron to Townsville, then Horne Island and Port Moresby. We got there to find they had lost a lot of aircraft. It was a rough strip but the Kittyhawk was a sturdy plane. However, if you strayed off the strip you were in real trouble.

Replacement aircraft were often flown in by ferry pilots. Bob Hartman comments:

When the Kittyhawks started coming through we took a lot up to New Guinea. I did five or six trips and brought old Kittyhawks back.

We'd normally stay at Amberley before a trip to New Guinea. We could have gone further but Amberley had the best mess in Australia. The catering officer was the former manager of the Hotel Australia in Melbourne. It had the best food on the coast.

We flew to Rockhampton to refuel, then to Townsville and Cooktown. We were advised not to go to Cairns. It was considered dangerous as the strip was not very long, there were mountains all round it and it had a right hand circuit. Pilots were not used to right-hand circuits.

We bypassed Cairns and went to Cooktown, then

on to Horne Island—right on the tip of Cape York before you get to Thursday Island—and then there was a two and a half hour trip across the water to Port Moresby.

Sometimes we varied the route. We might fly from Cooktown to Port Moresby with much more flying over water. Usually there would be five to ten of us for that trip. We were never allowed to fly across the water without having a navigator in a Beaufort or Beaufighter with us. We were armed on the first trip we did.

I did two trips in Vultee Vengeances—and then only because I had a wireless operator in the back seat. But even then I wasn't allowed to go direct to Moresby across the water. I had to go the long way round—across Thursday Island, past Duru, across the Bay of Papua and along the coast to Moresby. There was not much water to cross that way—you were hugging the coast all the way.

The weather was usually good, although I was once held up for a day at Horne Island. But once you got to New Guinea you didn't fly through clouds because they're full of rocks.

We were flying from Port Moresby to a place called Nadzab one day in September 1943. We left mid-morning and were approaching the ranges, which were about 5000 feet at that point, when the clouds came down. We couldn't get across to Nadzab so we turned back to Moresby. But we couldn't go back either as the clouds had come down behind us.

We flew round in circles for a while, then headed towards Milne Bay. About halfway between Port Moresby and Milne Bay we came across an emergency strip with a P-47 Thunderbolt left halfway down it.

We all had to land short but we all made it.

I was taking planes up there for three or four months until I went into the Test Flight in March 1944.

Most of the airfields in New Guinea were made with steel matting—one of the best inventions of the war. I reckon it just about won the war. But it rattled like crazy when you landed on it.

SOLOMON ISLANDS

Bougainville and the adjacent island of Buka were part of the Mandated Territory of New Guinea which was administered by Australia. They were geographically part of the Northern Solomons.

The Japanese began action to occupy these islands early in March 1942 and the tempo of invasion increased towards the end of the month. Some Allied coastwatchers remained in the area. Tulagi, further south, was subjected to air attacks at the beginning of May and was occupied on 3 May 1942.

This area was far away from Allied airfields but the crews of the Catalina squadrons tried to harass the enemy and successfully dropped supplies to the coastwatchers.

Catalina crews at this time consisted of a mixture of aircrew and ground staff musterings. The two pilots, a navigator and sometimes, a wireless operator/air gunner were aircrew musterings. The flight engineer and gunners (the latter were generally armourers) were ground staff personnel. This arrangement caused some dissatisfaction in the ground staff ranks.

RAAF activity in the area is described by Jim Cowan, who was a navigator with 11 Squadron. He tells of his stay in the area near Tulagi:

We had to disguise the fact that we had a Catalina station at Tulagi. The three Catalinas would take off at dawn. Two would proceed on their missions and the third would deploy (hide) for the day.

We broke the back of our Catalina while deploying one day. We were landing in a nice smooth lagoon when the skipper took it down to about thirty feet and just dropped it in.

A Catalina has a keel which consists of two members. If this gets twisted the plates are distorted and water comes in. The water poured in and we had to keep pumping while we jettisoned gear (including our bombs) to lighten the aircraft. We managed to get back to Tulagi that evening. They fuelled us up and we took off the next day, 11 April, for Rathmines and repairs.

Catalina. The glass blisters to the rear are gun positions.

When the squadron left Tulagi we gave most of our fuel to the mission station. I well remember the excited old Dutch missionary asking if his motors would run on it. We told him they would run very well if he mixed some kerosene with it. I hope they did.

The Catalina flights averaged fifteen to eighteen hours. You couldn't work the whole time so you had a bit of shut-eye and then carried on.

Later in the war we went from Bowen up to Buka and on 11 June 1943 I went from Cairns to Bougainville. That trip was eighteen hours and thirty-five minutes— six hours thirty-five minutes in daylight and twelve hours at night. We were dropping supplies to the coast-watchers. The trips were done on moonlit nights to give some visibility but were unpleasant because we had to come in low and the visibility was often not too good.

We were given clear map references and recognition signals for the drops but we never saw who we dropped it to.

Leo Allen was an armourer who served continuously as an air gunner on Catalinas at Bowen and Cairns from 22 August 1942 to 2 September 1943.

He describes his experiences on Catalinas and some sorties into the Solomon Islands area:
I was posted to Bowen on Catalinas. I had never seen one before. They were there because they had just lost four aircraft on the water at Port Moresby.

At first we only had Lewis machine-guns which had been used in World War I. We had them for the first four months. When the Yanks came we got .303 inch Browning which were much more advanced and had better gunsights. Later we got .5 inch Browning which were better still.

There were fighters as well as flak in the Solomons and we were saved many times by low cloud formations. There's always a lot of cloud over the Solomons and north of New Guinea. You could fly for fifty or a hundred miles before coming out. There were times when we flew very low over the water.

On 1 September 1942 we attacked Buka, an island off the northern tip of Bougainville. There was flak everywhere and we were well in range at 8000 feet. Despite the flak my pilot (Athol Warne) flew down the searchlight beam and we strafed the light as we came down.

Peter Fisken was stationed in Bougainville in the last phase of the war from May to August 1945:
I was on 5 Squadron on Wirraways and Boomerangs. It was army cooperation work in the mopping up of Japanese resistance.

We dropped flares on the Japanese troop concentrations as markers for the bombers and the army. Royal New Zealand Air Force Corsairs provided the bombing. They followed ten minutes after us.

I saw out the rest of the war there.

In the final stages the New Zealanders improved their map-reading techniques and the RAAF Wirraways were phased out of this activity.

Peter Fisken was still on Bougainville at war's end:
The day after peace was declared a mate and I, still half sozzled from jungle juice on the night before, climbed into a Wirraway and flew up to Buka, a jungle strip riddled with bomb craters. How he got it down I'll never know.

We got down and, to our surprise, these bloody Japs who were there didn't know the war was over. They took us to a colonel who was the only bloke who spoke English. He was in hospital with malaria. We had a hell of a job explaining to him that the war was all over.

They had been heavily bombed and had no radio contact but he finally got into contact with someone who confirmed our news.

NEW BRITAIN AND NEW IRELAND

These islands were occupied by the Japanese on 23 January 1941. Little opposition could be mounted.

In April 1943 it was decided that mines should be dropped in the New Britain and New Ireland areas. Mining had been successfully undertaken by the RAF in the European theatre of operations. The long-range Catalinas were called upon.

Jim Cowan:
We flew from Bowen to Moresby, landed there and then followed the coast.

Leo Allen commented:
Our planes were painted black to avoid detection. We dropped our mines at Silver Sound, New Ireland. We came down very low.

With the occupation of the Trobriand Island Group by the Allies on 24 June 1943 it became possible to harass New Britain with flights of Vultee Vengeances from Kiriwina Island. Ron Barker of 24 Squadron describes a sortie:
The Japanese occupied island of New Britain was 180 miles to the north of us. There was also a Japanese naval base at Yasmata. We had attacked troop concentrations and anti-aircraft guns there. This time we were to attack the naval base—and that didn't go down too well— particularly as there was no land at all between our base and the target.

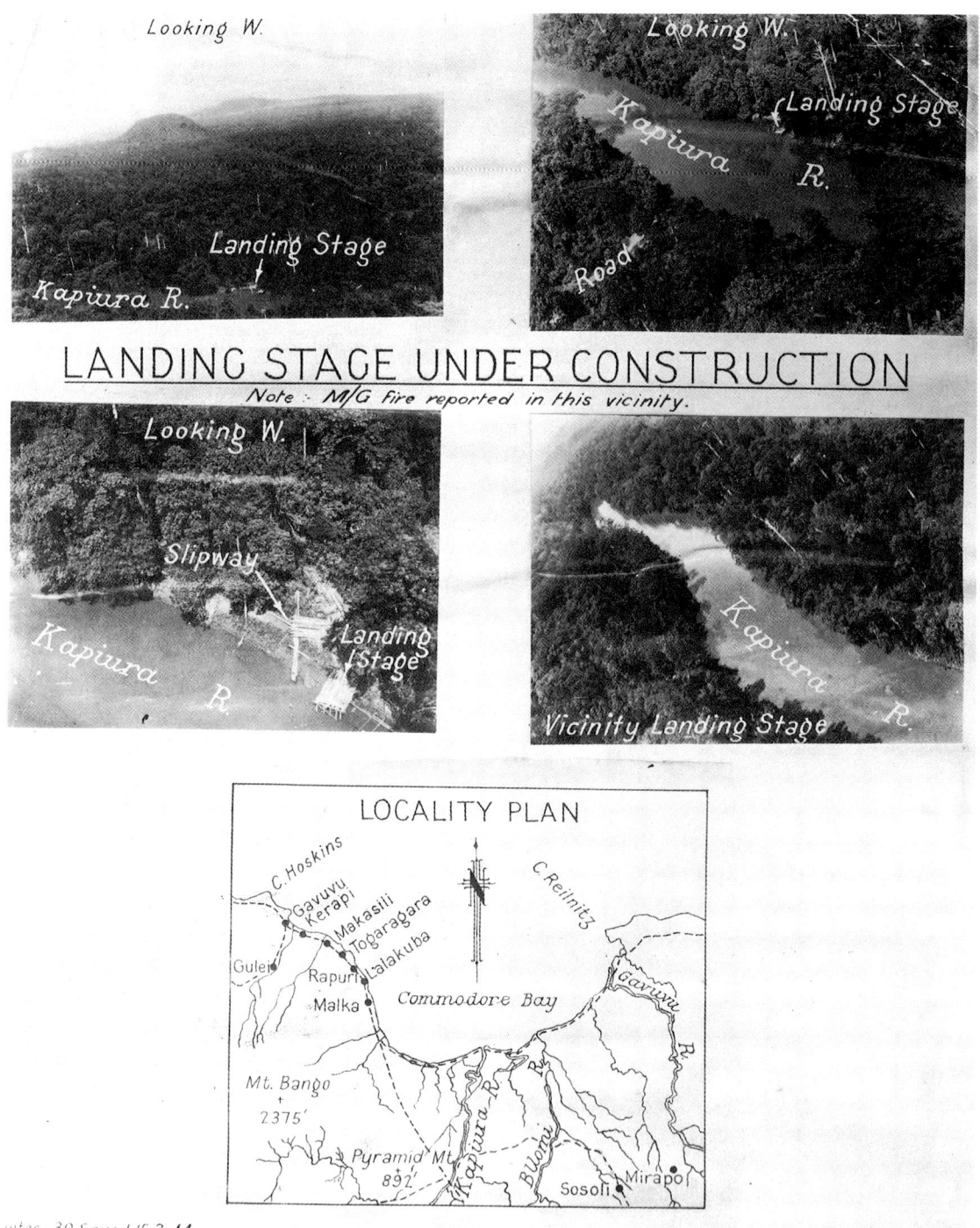

Briefing sheet for an attack on Landing Stage on Kapiura River, New Britain.

We set out in the morning, and my particular job was to take photographs of the anti-aircraft guns around the wharves. I had an F24 camera strapped round my neck and had to hang it over the side as we went down in a dive.

The F24 was made of an aluminium casting and was heavy to hold. I was hanging over the side and there was enough weight to hold me against the slipstream. I had to wind up and pull the trigger each time I took a photo.

There were supposed to be naval supply vessels there but we didn't see a great concentration of them.

We had engine trouble. We were blowing clouds of smoke all the way there and back. A Boston came up alongside and photographed us.

When we got back, one of the boys, who had been badly shot up over the target, crashed on the strip. He was in hospital for three months. That was one of the worst casualties we suffered. We never actually lost a man.

DUTCH NEW GUINEA

Noemfoor

This was a small island off the northern extremity of Geelvink Bay in Dutch New Guinea. It had three airstrips, in various stages of construction.

On 21 July 1944, an American force invaded the island. Enemy resistance was weak. Kamiri airfield was seized on 2 July, Kamorsan was occupied on 4 July.

Fred Cassidy, a wireless operator/navigator, operated from Noemfoor in August:

In August 1944 we went on a sortie to the Serang Islands. On the way back, after we'd been in the air some three and a half hours, I got a message that one of our Beaufighters was missing and we should conduct a sea search for him.

We did many miles of square search, and eventually arrived back at Noemfoor after six hours and ten minutes in the air—which is a fair time for a Beaufighter. We ran out of fuel as we taxied back to the bays. Unfortunately, we didn't find the plane.

Some days later, we had a message that the plane's crew were on Napier Island. No-one knew where it was—it was just a little dot.

These two guys, Smith and Gleeson, had landed on the beach. They didn't know where they were, so they searched around and rowed across to another little island. They still didn't know where they were. So they searched around again and found a further little island with a house on it.

In the house, they found a medicine bottle with a prescription for a Mrs Brown to take it three times a day. It carried the name of a chemist shop at some place in Queensland. So they returned to their plane and radioed all this back. After all this, it was eventually found that Mrs Brown lived on Napier Island. So we went to Napier Island to see what we could find.

Smith and Gleeson slept on the beach that night. While they were there, a party of Japanese came across from another island. They climbed all over the Beaufighter, but merely looked. They didn't disable it. They simply went back to where they came from. It frightened Hell out of Smith and Gleeson—they only had Smith & Wesson revolvers.

Eventually, a P/T boat came out with spares and a crew of engineers to see what they could do to get the plane airborne. Smith reckoned that, at low tide, he could get it off. He'd got it down, so he could get it up! The beach was only about 500 yards long.

There were about ten Marines on the P/T boat. Three of them rowed to the shore of another island and

disappeared for a while. Then we noticed one of them on the beach, crawling along and yelling out something or other. We kept strafing to give some support, but a Japanese came out and shot this fellow on the beach. I found out later that the other two had been shot and killed. The other marines made it through 100 yards of water up to their armpits, with their rifles over their heads.

The engineers finally got the Beaufighter working and Smith took it off.

Jim McSharry joined 22 Squadron at Noemfoor in November 1944:

I was posted to 22 Squadron, which was a Boston squadron, at Noemfoor. We had trained on Beaufighters and Beauforts, but I had flown Bostons as ferry pilot and was quite familiar with them. They had a tricycle undercarriage with a nosewheel.

Our job was bombing. We bombed Japanese barges and, sometimes Japanese aircraft on airfields. There were some Jap aircraft in the area, but we met very little opposition from them.

Our main problem was flak. The Japs had quite a bit of anti-aircraft on some of the more heavily defended targets and they were pretty good shots.

We would attack from a few hundred feet. I can recall only one raid when we were briefed to attack from 5000 feet.

Casualties were fairly light.

Morotai

Morotai Island was situated halfway between New Guinea and Mindanao, in the Philippines. Its capture would provide the opportunity to give air cover for the advance towards MacArthur's goal—the recapture of the Philippines.

On 10 September 1943, Morotai was attacked by a task force which encountered no opposition on landing. The weakness of the Japanese resistance allowed the 61st Australian Airfield Construction Wing, including No 14 Airfield Construction Squadron, to commence repairing Wamma airfield on 24 September. It was open for reception of aircraft on 4 October.

Squadrons, or sections of squadrons, were outposted from Morotai on occasions. One such occasion involved Jim McSharry:

I became friends with P/O Jack Chamberlain on the squadron. He had been in Training Command, towing drogues, etc. He'd gone on to 30 Squadron, but 22 Squadron was short, so two or three crews came over from Noemfoor. Our wing was 77 Wing RAAF and was made up of 22, 30 and 31 Squadrons.

Late in the war, the army decided they were going to take an island called Labuan, off the coast of Borneo. I was detailed to take half the squadron to Sanga-Sanga.

One day, we were on an army support operation. We were dive-bombing, and a battalion of AIF were trying to drive the Japanese off the island of Labuan. We were required to make the short trip from Sanga-Sanga and, as we had plenty of fuel, circle for a while until we were directed to a target.

There was a controller on a ship in the bay. The army would notify him of a problem. If he felt that the aircraft could help, he would call up the leader of the aircraft and give him the target to be bombed.

The Beaufighter was a good aircraft in which to dive-bomb, and we did quite a bit. This time, the target was a Japanese pillbox—a big, fortified, foxhole which was being used as a machine-gun post. It was harassing the AIF, who were reluctant to charge it if there was some other way of getting rid of it. The CO called the controller, who called me, and gave me a map-reading.

I was leading the group of six, and my friend Jack Chamberlain was leading the second three. I sent him in first—showed him the target, and said, 'See what you can do.'

Jack went in but wasn't satisfied with his angle of approach. He aborted his dive and sent his No 2 in. The bombs landed all round the post, but didn't hit it. It was the sort of thing that required a direct hit.

After Jack's aborted dive, I went in. I missed also, as did my other two blokes. I called Jack and said, 'It's up to you.'

He went down in a perfect dive, and let his bombs go at the very last moment, to be absolutely certain. I thought for a second that he wasn't going to get out, but he did, and he dropped his bombs right into the pillbox.

When we got back there was a message from the colonel commanding—'Thanks. You saved a lot of Australian lives.'

Jim McSharry was involved in another incident which illustrates the vagaries of the weather in the tropics:
One day, the whole squadron was stood down. I happened to be in the vicinity of the Ops Room when an emergency call came through to say that a Kittyhawk had ditched at sea. They wanted some aircraft to go and search for it.

The pilot had ditched north of the Celebes. My navigator wasn't around, so I took the aircraft and navigator of my friend, Jack Chamberlain. We went off, half a dozen of us, to search for the Kittyhawk.

Unfortunately, we didn't find him. We searched until dark, and then had to return. There was no point in searching any further.

On the way back, we ran into a terrific thunderstorm. When we got into this cloud, we had no idea of exactly where we were. My navigator was trying to get a bearing from the airport at Morotai. He couldn't get through and our fuel was getting pretty low—things were beginning to look very very grim.

All of a sudden, over the R/T, I heard the voice of another of the pilots. He said, 'This is Mitch. Fly south. You're north of the field. You're way off. Fly 180 degrees.'

So, I turned south and flew for about ten minutes. I came out of the cloud and saw the lights of the airstrip. Down we came.

BORNEO

MacArthur considered it necessary to invade Borneo despite the fact that Britain did not have any wartime need for Brunei and Blamey, the Australian commander, did not want to move in the manner prescribed.

Tarakan

The campaign was launched and the 26th Brigade of 9th Australian Division landed at Tarakan (Dutch Borneo) on 1 May 1945. It was intended to build aerodromes there as quickly as possible to support the attack on Balikpapen. The Japanese were quickly driven inland.

Bob Hartman was there soon after the landing:
I was posted to a Repair and Salvage unit (RSU). The function of an RSU was to do repairs and maintenance for squadrons who then had only to do routine day to day servicing. Anything major, such as a major overhaul or changing an airscrew or a mainplane, was done at an RSU.

I was the only pilot. The unit was otherwise staffed completely with engineering people. The working conditions were primitive. All work was done out in the open with only a tent fly over the top to keep the sun off.

As a safeguard against malaria everyone had to wear long trousers, long-sleeved shirts and canvas puttees after sunset. Malaria had been quite common until they introduced this precaution. The men working on the machines could take off their shirts in the daytime because of the heat but had to put them on and cover up as soon as they knocked off.

Bob Hartman in tropical rig with a Kittyhawk.

Some of the men had been overseas for over two years and were ready to go home. When I joined there was a new CO and he was much stricter than the old one. This created some friction and they were very very browned off.

The airmen were allowed two bottles of beer a week plus a cigarette and tobacco ration. The officers were allowed two bottles of beer and a bottle of spirits per week. We pooled all our beer and spirits and were able to have a few drinks each night.

The airfield construction units went in with the troops at Tarakan with the idea of having the airfield ready in a few weeks. But they found it immensely difficult to build an airstrip. There was no rock to be found. It was an oil-producing island and, as far as they could make out, it was just a lump of mud floating on a sea of oil.

There were torrential falls of rain at night which washed out all the work done on the day before. There were occasions when a truck left out overnight disappeared in the mud by morning.

An advance party of our unit went in a couple of days after the army and set up camp. They managed to get hold of an old Dutch house which became the Unit HQ,

Officers' Mess and officers' quarters. Most of us had a room with a bed in the house. The rest had to live in tents.

I went across about fifteen days after the landing. Another officer and I went in an American LST with truck after truck of equipment which we just drove on. We shared the officers' quarters with the ship's officers. They were all white and the crew were all black.

The officers treated the negroes with complete arrogance—just like dirt. If it comes to that, they didn't treat the two of us with much courtesy either. They were contemptuous of everyone. They seemed to think, 'We are Americans and the rest of you can go to Hell.'

We were at sea two days and one night. The sea was flat and there were flying fishes playing. The food was good. When we arrived we just drove the trucks off.

As the strip wasn't finished I just sat on my bum for a few weeks. The CO had me doing all sorts of jobs— education officer, barracks officer, sports officer— anything to keep me out of mischief.

When the strip was finished 75, 78 and 80 Squadrons came in and I got to work. I was in charge of the test flight and aircraft pool. I carried all the spare aircraft that were available to go back to the squadrons as replacements.

When an aircraft had undergone an overhaul I tested it. When we started I hadn't flown for two months and hadn't flown a Kittyhawk for eight months.

The first aircraft I took up for test vibrated very badly. When I got down I said to the Engineering Officer, 'I think the airscrew is out of alignment.' I was right. When they found I knew what I was talking about I was held in much more respect.

Labuan

Labuan Island is situated off the north western coast of Borneo (Kalimantan). It commands the entrance to Brunei Bay.

The township of Labuan provided port facilities at Victoria Harbour.

Taken by the Japanese in 1941, it was attacked by the Australian forces on 10 June 1945. The 2/43rd and 2/28th Battalions landed unopposed and immediately set out to capture the airfield and that was done the same evening.

The airfield became operational on 17 June 1945. Bruce Edenborough, who was attached to No 9 Local Air Support Unit, operated from here on a variety of tasks.

He comments:

We flew Beaufreighters—they had taken the turret out of the top and made a gradual slope down to the tail. Our unit operated from Labuan just after it was taken. We had all kinds of places we had to go in and spray— such as Limbang and Totar—as an anti-malarial precaution. We would go in at treetop level and spray the whole area. When we got back to base we would do our own unit's area with any remaining DDT.

One of our main tasks was to drop food supplies to a group of people called SRD (Special Reconnaissance Detachment) who were people who operated wireless equipment behind Japanese lines. They also supported the Dyaks who used their blowpipes to put darts into the Japanese wherever they could find them.

This was very interesting because of the places we went to in Japanese-held territory. Places like Bawang, Tenhillan, Tengla, Iburn and Miaow—although there were problems in finding these places. We would locate them and drop food canisters containing sugar, flour and tea. We also dropped kerosene or petrol for the motors to recharge their batteries.

I remember we once dropped in a very narrow valley with a big river running down it. There was a bend in the river with two big long houses on the bank. One canister landed alongside the long houses but the other landed in the river. We had to go back and drop another canister.

Then we made all sorts of trips around the place on various duties until we heard, one day, that the atom bomb had been dropped in Japan and that the Japanese had capitulated.

That gave us another big job. The Japanese-held territories were far-flung and many local commanders couldn't believe that the Japanese Army could surrender. We had to go into all kinds of places up and down the coast dropping leaflets telling them that it was all over.

Kuching was slow to capitulate and HQ had heard that there were Australian soldiers in a prison camp there. They sent us down with a big escort of Mosquitoes to locate the camp. We flew round for some time looking for it and then I received a message from the Mosquitoes that they had found it. We then flew round just above treetop level taking photos of the camp until petrol got low and we flew back to Labuan.

The photos showed it was full of POWs so they sent down a navy ship which brought back a lot of men. They ferried the rest back to a big camp in Labuan.

We made quite a few trips down there to take in personnel and help bring men back. Our first trip was to see what the strip was like. The Japs reckoned they had fixed the many bomb craters in it. It wasn't a big field and when we touched down we ran over the edge of one of the filled-in craters. I felt the tendency to pull the aircraft to the left but could control it so everything was well. We made many trips down there then.

The men brought back were very emaciated—just skeletons in many cases. One of them ate some eggs, I don't know how many, but lost them all because his stomach couldn't handle the intake of food. They had to be brought back very slowly to a full diet.

TIMOR

On 27 November 1942 plans were made to evacuate Dutch troops (who were to be relieved), some Portuguese refugees and 2/2nd Independent Company from Timor. The corvettes HMAS Castlemaine and Armidale were to provide the escort.

HMAS Armidale was sunk by Japanese early in December. Leo Allen was flying in a Catalina and was involved in the search for survivors. It was not considered safe to base Catalinas at Darwin at the time so he flew from Bowen to Cairns and then to Darwin:

One of our naval ships was sunk in the Arafura Sea in December 1942. We did some sweeps and saw rafts but couldn't pick up survivors as the waves were as big as houses. We dropped Mae Wests and rafts. We went out three times but only saw them once.

Forty-three of the crew were finally rescued but forty lost their lives.

Labuan Air Force Camp.

HELP TO ALLIED AIRMEN.   B.M.17

     The bearer of this letter is the pilot of an Allied plane which has crashed. You must help him as best you can. You should give him food and water, and if he is wounded you must look after him until his wounds are healed.

     Moreover, hide him from the Japs and help him to return to their own forces.

REMEMBER. This man is a friend who is helping to throw out the enemy from your country.

     When the Government returns you will be rewarded for any help you have given to Allied airmen.

REMEMBER. As long as the Japs remain in your country you will be poor.

     But when the true Government returns, peace and happiness will return.

---oOo---

Bruce Edenborough dropped this pamphlet to the Japanese advising them of the Emperor's surrender. He carried the the chit on the right to give to locals after a crash.

Bruce Edenborough drops supplies behind enemy lines in Borneo, parachutes at top middle of the picture.

Kuching P.O.W. Camp.

5

THEY ALSO SERVED

Ground staff were the too often unpraised workers of the air force. Yet, preparing meals, servicing aircraft, driving transport, fitting cameras, manning aircraft control, or any of the other many activities of the men and women on the ground were all vital elements in the operational effort. It is worth noting that Sir Arthur Harris, **The Bomber Offensive,** *points out that, in his command, some 8000 men and women of ground staff categories were killed in the UK, 'in training, in handling vast quantities of bombs under the most dangerous conditions . . .'*

Ground crew served in all manner of tasks, in areas all round the world—in Iceland, in the Western Desert, in the United Kingdom, in India, in Burma, and in countless other areas.

My stories are few, but do, I believe, reflect on the extraordinary variety of ground crew tasks and experiences.

UNITED KINGDOM

Ernest Blowman tells of one hazardous occupation on 462/466 RAAF Squadrons at Driffield, Yorkshire:
I was collared to be Defence Officer's Clerk. It was here that I met these lads. One was on bomb disposal—and we had a few to dispose of early on.

His glengarry (cap) was stiff with oil and grease, he had a scruffy notebook, a pair of pliers and a screwdriver. His name I do not recall, but he used to contact me every morning for any new German bomb fuses that had been discovered.

I remember one bomb that had dropped straight down by a hangar door and was in a right awkward place to get at. No trouble. He soon defused it and a tractor towed it away.

Jim Savage, fitter IIE/air gunner with 10 and 461 Squadrons, gives another example of the devotion to duty and the complexity of the characters of those engaged in ground duties:
If you were asked to describe 'Ozzie' Ferguson the first word to come to mind would be 'rough.' Not just rough but 'rough as guts.' He appeared to have a limited vocabulary; fully 90 per cent of his utterances consisted of old Anglo-Saxon four-letter words with variations beyond the imagination of old-time wharfies and bullock drivers.

I first met him when he was a corporal fitter IIA (or rigger) in the hangar at Mount Batten with a gang who were not known for their gentility or moderation. They did not ape Oz, they were individuals in their own right.

As Oz progressed in rank to flight sergeant in charge of the riggers there was little or no change in his manner. He was just as rough as ever but everyone in Maintenance, from the Engineer Officer down to the lowest aircraft hand, thought he was just great. He simply had a way of getting things done and was quite prepared to work the long hours he asked of his men. There is no doubt he was a great leader.

Whilst he had flown as crew in the squadron's early days, the long hours out over the Atlantic and the Bay of Biscay (they were quiet spots in '39-'40) bored him and he hurried back to Maintenance Flight.

Funnily enough he carried a disdain of all those of us who had got into Operations Flight to fly as fitter IIA/air gunners even though that is precisely what the brass had in mind for us when we were chosen back in Australia for 10 Squadron. Oz considered us lazy bastards who flew on ops to escape the rigours of working in the hangar. He gave us hell when our aircraft had to go up on deck for inspection.

As soon as the beaching gear was attached and the winch cable started the aircraft up the slipway he would be aboard. He looked inside all the storage areas and pulled up all the floorboards to check the bilges. His gimlet eyes missed nothing.

He once made everyone from the first fitter (afterwards flight engineer) to the tail gunner lose a day's leave. We were all ready to leave the aircraft and bolt for North Road Station en route to London for fun and games when Oz came down on us hard. We had to don overalls and, with much soap and water, literally scrub the bilges clean. All this was because some silly ass had plucked a chook—which had got lost up at Angle Bay at Milford Haven—and let the feathers fall down into the bilges. We were innocent and prepared to swear it but it didn't make any difference to Oz. His 'boys' were not going to clean up that kind of mess. My back ached for days.

However, we were revenged the next time we came up on deck. when the aircraft came to rest at the top of the slipway it had to be manhandled on the tarmac for there was not enough space for a tractor to manoeuvre its thirty tons. Only manpower pulling on port and starboard ropes could do the job safely. It was bitterly cold with some sleet and there was ice in the puddles. No matter how thirty to forty men tugged, the aircraft would just not go where Oz wanted it.

He was standing on the port side when he spewed out a string of beautifully composed obscenities reflecting on the sexual proclivities, ancestry (back three or four generations), ape-like appearance and general uselessness of the hauliers. All this was at the top of his voice. It really was choice.

Just then a young and innocent WAAF officer (yes, they did exist) and a little corporal appeared at the nose of the aircraft, probably on their way to the ops room. With one accord the boys dropped the ropes, sprang to attention and, with all the power of their young Aussie throats roared out, 'Flight Sergeant Ferguson is the name Ma'am.'

Poor Oz was stunned and stood with mouth agape as the boys, with renewed strength, picked up the ropes and pulled X for X-Ray almost into the hangar.

Oz recovered but was known, from then on, to have a quick look round before letting fly.

One morning in autumn 1944 a special muster parade was called and the whole squadron, except those in the air, stood waiting for confirmation of a rumour which had raced through the squadron the night before.

The CO came out in front of us and, without preamble, told us that Oz had 'bought it.' This is the only time I saw the CO cry—well his eyes were rather wet.

Oz had worked himself (and his gangs) into the ground fitting four Browning machine-guns up forward as sub-strafers to save our aircraft from being shot up by the extra guns now mounted on the U-boats. Even then it took an MO's order before he would take leave.

He was standing in Kingsway talking to Jack Forbes when a V-1 (buzz bomb) came down near the Air Ministry. He suddenly collapsed at Jack's feet. It only took one lousy little piece of shrapnel in the back of his neck to still that rasping roar forever and make not just a CO cry but almost a whole squadron.

Les Mills was an electrician in the RAF. He served in Great Britain, including the Shetland Islands. He explains the reason for one of his postings:
I was stationed at Crosby on Eden, near Carlisle.

I was fixing some signal installation to the roof of a hut under the direction of the Sergeant Electrician (a Geordie). He was difficult to understand and I was on the end of a taut line of piano wire used for bracing. I thought he said, 'Let go!' when, in fact, he said, 'Fasten it!' The result was that the wire sprang back and he ended up with the end of it stuck in his head. It was only extracted with difficulty.

For some reason I was posted a week later to Wick, a remote airfield in the north of Scotland. It was in January and the weather was so inclement (four feet of snow) that it took me four days to get there. The train couldn't get through the snow and had to keep stopping to get it cleared.

Although he was an electrician, Les was also employed as a driver—not always a safe job:
One of my jobs was to drive the tractor which pulled the mine trolleys taking mines back to the bomb security area from planes which had aborted.

One day I was driving and looking back over my shoulder every now and then to make sure my load was secure. Suddenly, when I looked back, I was greeted by the sight of an empty trolley, no mines and no guard who was supposed to be ensuring their security. There were mines scattered all over the perimeter track. We had to find them and roll them back on again.

On another occasion a plane being refuelled had a load of depth charges on it.

A daily inspection was being done at the same time. I looked out of the window of the flight office. Refuelling was being done from a tanker but this was now shooting across the drome and everyone else was running away as fast as they could.

Apparently there had been an electrical earth fault while testing the undercarriage. This resulted in the depth charges dropping from the bomb racks to the ground.

Of course, not everything in the garden was lovely. Jack Stronach, for example, tells of one time when the great degree of care and attention which should have been applied, was not. His story relates to the duties of the armourers, who were responsible for bombing up and for servicing and testing the guns:

On 16 May 1942 I was stationed at No1 Ferry Training Unit at Lyneham in England. We were carrying out various air tests in preparation for ferrying a Hudson aircraft to India.

I was to carry out a fuel consumption test by flying to the Irish Sea, round the top of Ireland, out into the Atlantic and return—a trip of about seven and a half hours. I had taxied the aircraft over to the control tower, which had the station commander's office on the ground floor. We were waiting for permission from Fighter Command before we took off.

I was sitting, facing the rear of the aircraft, on the main spar of the wing, which passes through the body of the machine. I had my elbows resting on my knees and my head resting on my hands as I contemplated the floor.

The door of the aircraft was opened and an armourer put his head in and said, 'Good morning, sarge.' I looked up, said, 'Good morning,' and put my head down again.

I didn't notice that the armourer then entered the aircraft with a second armourer and took the Verey pistol from its holder, which was attached to the side of the aircraft, opposite the door. The pistol was loaded with the colours of the day.

I heard the first armourer say, 'Have you seen the new safety catch?' I looked up and he pulled the trigger of the pistol. It fired and the light came straight at me. I leaned to the left and raised my right arm. The Verey light passed under my armpit, hit the wall behind me, ricocheted around the walls and came to rest under a long-range fuel tank

The aircraft soon filled with smoke and my navigator, Eric Ball, who had been in the navigator's position in the nose, came rushing past me to get out, having no idea what had occurred. I grabbed the fire extinguisher near the door and went back to the fuel tank, which was full, and extinguished the Verey light.

People rushed out of the control tower to see what was happening as smoke poured from the aircraft. On examining my battle jacket later, I found scorch marks and bits of metal in the front.

John Hopley's story is set against the background of the invasion of Normandy in 1944. The Tactical Air Force was formed to give close support to the invasion forces as they moved into France:

All aircraft had been refuelled and rearmed in readiness for the next operation of the day, and I was reading in my tent, the airmens' tents having been pitched in the dispersal area, close to the aircraft. Suddenly, the relative peace and quiet of the afternoon was broken by a commotion outside, which resolved itself into the voices of the NCOs of the unit, running through the tent area, shouting, 'Everyone out—at once—let's have you—two-six on the kites' and other urgent-sounding cries and threats.

On leaving my tent to join in whatever was happening, I became immediately aware of a large cloud of black, oily smoke, which was coming from the general direction of the tent site, and which was being blown towards the petrol bowser, and, beyond that, to the aircraft.

Someone drove the bowser out of the way while the airmen manhandled the Spits out of the range of the smoke, which by this time, had burning fragments in it. In a few minutes, the Wing fire tender arrived and put the fire out with foam.

The reason for all this activity can be traced to Charlie Jones.

It was Charlie's day off. He had been on duty crew the previous night. His tent mates, members of the same crew, and also on day off, had gone into Chichester but Charlie stayed on the airfield to do his laundry. To do this he needed hot water, and this was usually obtained by boiling up a bucket, or other suitable container, over a fire produced by soaking a tin of soil or sand with a cupful of petrol, usually obtained from the bowser driver, and applying a lighted match. The result was a fire of considerable heat—if somewhat smoky.

As this was an advanced landing ground, there was no piped water, but a water tanker visited each site daily, and a supply was obtained by filling all available receptacles, which often included empty petrol tins.

Having got his fire going, Charlie put a petrol tin of water on it. However, he didn't know—until a few minutes later, when it became quite apparent—that his tin of water was, in fact, a tin of 100 octane aviation fuel. It exploded, and the wind blew the flames towards Charlie's tent. It immediately caught fire, together with its contents—clothes, uniforms, blankets, rifle butts and other personal kit. It was at this stage that the alarm was given, as related above.

Charlie was put on a charge, of course, but the Wing was moving shortly to Normandy, and it was not until they arrived there that the system caught up with him, and his charge was heard. He was fined one pound and awarded seven days' confinement to camp.

However, with the aid of a sympathetic flight sergeant, Charlie, a flight mechanic (engines)

'volunteered' for duty crew each day of his sentence. While this meant that he put in a considerable total of working hours, he was spared the indignity and inconvenience of having to report to the guard tent at frequent intervals and being put on fatigues in the evenings. Charlie reckoned this was a reasonable trade-off.

MIDDLE EAST & MEDITERRANEAN

That an army marches on its stomach is a trite saying from periods past. But, so too does an air force. This was particularly true in the Middle East, where food was basic and unimaginative. Any attempt to improve things worked wonders for morale.

Let's look at the views of Ron McCathie and Len Barton. First,

Ron McCathie:

LAC Johnson was a lovely Pommy bloke, about forty years of age—loyalty was his middle name. I 'won' him as my 'dog's body' on becoming Messing Officer for 1438 Stat. Rec. flight.

During the time the unit was operational, he and I scrounged tucker for our mess, all the way to Persia and back to Egypt.

Using a fifteen-hundredweight truck, our foraging took us into some risky areas among the fiercely independent Kurdish tribesmen on the Turkish border with Syria and Iraq. We looked for eggs, chooks and anything else. We never wore sidearms, but carried lots of Wm Angliss bully beef as our 'currency'. Once the natives were convinced it was not pig's meat, we had no trouble. We could get four dozen eggs for one tin of our Australian product. Chooks, ducks and pigeons came a little dearer. Vegetables were very cheap.

Some of our journeys into the mountains covered fifty miles or more, and there was always a great deal of amusement when we returned with our assortment of farmyard collectables. Thanks to Johnno and the cook, we were well provided for. If we stayed long enough at the same airstrip, the villagers in the Kurdish mountains came to know us very well. This made trading much more fun, and a great deal easier.

The young men were never very visible. When you did see one, he always carried a rifle, with a bandolier of cartridges criss-crossed over his shoulders, and a dagger in a scabbard by his side. The Turks to their north were neutral and we, on their southern flank, left them alone. We did not wish to jeopardise the flanks of the east-west lines of communication between Egypt and India.

You would really have to traverse these mountain tracks, as we did, to realise how difficult it would be to dislodge them from their virtually impregnable positions on the mountain trails.

Reverting to Johnno. I felt we were indebted to him for his contribution to our comfort. The only way I could record our appreciation was to seek his promotion. Arriving in Tehran, I set about doing so, and applied for his corporal's stripes to HQ, with our CO's signature supporting the application.

The reply was negative. 'No establishment vacancies in our command in his category.' I persisted, and after appealing and reapplying on three occasions, over a period of three months, the Air Force bureaucrats at HQ must have thought me a pest, and disposed of me by 'promoting' John to lance-corporal, acting, unpaid. Have you ever heard anything like it?

Well, our friend could, at least, wear the much coveted stripe—and perhaps he may have been in front of the queue for confirmation of his promotion later on. I never found out.

Len Barton:

458 Squadron RAAF was formed in Yorkshire during October 1941. After operating with Wellington bombers, with heavy casualties, for some months, the squadron was ordered to the Middle East.

The aircrew had to fly their aircraft out via Malta, and the ground staff had to go by ship, via Durban, Mombasa and Aden. This move took place in February 1942. When the ground staff eventually arrived at Suez in May, they found that the commanding officer, and all the key aircrews and their aircraft, had been shot down. The remaining aircrews, and their aircraft, had been assigned to other squadrons. Thus 458 Squadron had ceased to exist.

While contemplating this situation, and awaiting a decision on their future, the ground staff occupied Fayid aerodrome on the Suez Canal. In the meantime, the British forces had retreated to El Alamein and there was much panic in the rear echelons.

In the midst of all this uncertainty and chaos, some strange, slab-sided aircraft with white star markings began to land at Fayid. There was much speculation as to their origin. The common conclusion was that they were Russian. However, this was quickly dispelled when men with strange helmets, Colt .45s strapped to their legs, and chewing cigars emerged from these strange machines and said 'Say, buddy, where can we get some gas?'— Yanks!!

It was the first USAAF unit to arrive in the Middle East. The planes were B-24's (Liberators). The unit was designated as Halvorsen's Project (Halpro), en route to China, to bomb Japan.

They stayed a few days, getting their aircraft serviced, and then, as the Middle East situation looked grim, offered to fly out some of 458 Squadron to Palestine. The men selected were mostly 'tarmac musterings'.

On arrival at Lydda, however, the onward movement of Halpro was halted and the RAAF personnel were officially attached to it. I had been sent with the detachment as the Disciplinary NCO but, when we got to Lydda, I was asked to act as caterer—before enlistment I was in the merchant service as a marine steward and pantryman.

We had landed in the late afternoon, and the only rations we had were twelve cases of bully beef and six cases of canned tomatoes. I managed to locate and buy a sack of onions in a nearby village and, with great improvisation, managed to turn out an evening meal for 200 men.

We found that the Americans had no arrangements with the British for drawing rations, but each man was given a daily field allowance of $6 for sustenance. I was authorised to take whatever amount of this I felt necessary to feed them. I collected this every morning on the breakfast chow line.

Things gradually improved, until I had a big stone mess hall, a kitchen with a large cool room, twenty native cooks and two RAF airmen, who were my right hand. I was buying cattle on the hoof in Syria, trucking it to Lydda and doing my own slaughtering. I bought the entire contents of the AIF Canteen Depot (including 160 dozen of Australian beer), when they departed for Australia.

The Halpro aircraft were Liberators and their main targets were the Ploesti oilfields in Rumania. Other targets were Tobruk, Benghazi and Tripoli. All the personnel of Halpro were regular US Army Air Corps with a high standard of discipline. At first, they were nonplussed at the relaxed Australian discipline and mode of dress (boots, shorts and hat), but they soon realised that the job was always done properly and efficiently, and the dress didn't matter. The officers regarded the Aussies very highly.

One day, when on inspection, the CO discovered the beer in the cool room and, despite my voiced misgivings, ordered that it be issued out to the Australians—alcohol was not allowed on a US base. It took days to round them up after that, but the Yanks forgave them.

When Colonel Halvorsen left the unit, he sent a Letter of Commendation on my work to RAF HQ, Middle East.

About that time, General Louis Brereton arrived from India with nine Flying Fortresses (B-17s) and the unit was redesignated as the 1st Provisional Bomb Group. Later, in August, elements of 98 Bomb Group (B-24's i.e., Liberators), and of the 12th Bomb Group (B-25s, i.e., Mitchells), arrived and were consolidated as the 376th Bomb Group. This brought the messing strength up to about 1000 men.

When things settled down a bit, I managed to get an occasional day off in Tel Aviv for a swim, and a meal of different food at the Cafe Plitz.

'Buster', a photographer, was one of the lads in the detachment. A big, heavily built fellow with a perpetual grin on his moon face, his hobby was beer and photographing the ladies of the local brothels, and their charms—particularly the latter.

One night in August I saw Buster in Tel Aviv and we agreed to share a cab back to Lydda. As we arrived at the drome, a B-24 attempting to land, hit a two-storey building which was being used as a barracks for Palestinian policemen. He and I rescued several badly burnt men from the ensuing conflagration and exploding ammunition. I was burnt on the ankles and hit on the leg by an almost-spent incendiary bullet.

I went to the Medical Section next day for some dressings. The Medical Officer told me that Buster and I were to be recommended for a medal. However, shortly after, Buster, who enjoyed a practical joke, blotted his copybook by sending some of his most lurid brothel photos, instead of the Ops photos, to the CO. Consequently, I got a medal (the Soldier's Medal) and he didn't.

We remained with the Yanks for some months, and when we left I was again Commended in Orders of the 376th Bomb Group for my work as Mess Sergeant. This latter produced the Army Commendation Medal.

About 100 members of 458 Squadron served with the Yanks, and I'm sure we all had a lasting impression of their good discipline and generosity. The US officers were loud in their praise of the Aussies and fought hard to retain us. But Cairo decreed otherwise and, in mid-September, we returned to Fayid. Shortly after, I was appointed to the staff of the Command Catering Adviser; but that is another story.

For many years after the war, Buster and I would meet on Anzac Day and drink to our medal. He's dead now, but I often wonder what became of his brothel photo collection.

Early in 1942, the Allied Command felt that the Germans might break through in the Caucasus and come down, through Turkey, into Syria. The British '9th Army' was created.

Len Barton served with the RAF in this decoy operation, which he describes:

After attending a dance in Abassier Barracks, three of us sergeants had an altercation with three MPs. Words led to a donnybrook, which we won, but they brought up reinforcements and we were heavily defeated, thrown into the guard-house and thoroughly chastised. After some hours, we were released and threatened with further legal action.

I cogitated on this for a few days, and decided that a change of scenery would be beneficial. So, using the Old Boy Line that operated between senior NCOs around Cairo, I had myself posted to a small unit proceeding to northern Syria as part of the mythical 9th Army. It was a wonderful trip, taking several days on the road, with magnificent scenery by day, and camping out at night.

Arriving at our destination, some miles north of Aleppo, we spent several days establishing a camp. Then, one day, I was ordered to parade, in best blues, before the AOC Levant in Aleppo. 'What for?'—'You'll find out when you get there.' I duly arrived and marched in to see the Great Man.

Without any preliminaries, he started reading out charge after charge laid by the Cairo MPs, alleging that I had almost caused the downfall of the Allied war effort. When he had finished, he looked up and said, 'How would you like to be Station Warrant Officer of RAF Station Aleppo?' Taken aback, I said, 'I would like that very much, but I am only a sergeant.' He said, 'It is in my power to promote you to the substantive rank of flight sergeant, and make you an acting warrant officer. Will you take the job?' I said, 'Yes Sir—but what has this to do with what you just read out?' He answered, 'We both know that MPs are liars, but if you can do one quarter of what they accuse you of—you'd make a damn good Station Warrant Officer!' And that's how I got promoted in the field.

As part of the mythical 9th Army, RAF Station Aleppo, in a matter of a couple of months, expanded from 150 men to more than 5000. The core of the 9th Army was an Indian Brigade, and the rest were odds and sods dredged up from around the Middle East. By ringing the changes with unit signs, it appeared to be several divisions.

The air force component had a Spitfire squadron and several squadrons of dummy aircraft as well as armoured cars from Iraq and various units of the RAF Regiment.

The Senior Administrative Officer (Squadron Leader John Tonkin) and I were the only Australians in the whole outfit.

The purpose of the deception was to deter the Germans from coming down through Turkey. After about four months, the 9th Army just melted away. But I kept my f/sgt rank. I was later promoted to warrant officer by more orthodox means.

Walter Mould describes his experiences on Malta:

After arriving I went to Luqa, the main airfield, where I was to spend the rest of my time on Malta.

Malta has been described as the land of bells, wells and smells. Water, or the lack of it, was a problem. The island is mainly limestone and there are no permanently running streams. The people who originally settled the island in remote times are unknown but the Carthaginians settled there later and their legacy is the basically Arabic language.

I was lucky to be billeted at Siggiewi, the radio station. There the barracks were large and airy, built in Mediterranean style. The underground shelters were hewn out of the rock. There was a canteen and, for a while, life was good.

Siggiewi was situated about one and a half kilometres from the edge of an airfield but it was three kilometres from where I worked most of the time. Sometimes we walked back for lunch. Usually we used a bus to go to work.

Work consisted of dealing with aircraft in transit to Egypt, Western Desert, Iraq, India and further east and, some, to Gibraltar. Wellingtons and Bristol Blenheims were based on the airfield.

There were three other airfields, La Kali (or T'Kali), Hal-Far and Kalafrana. They were fighter, Fleet Air Arm and flying boat bases respectively, through rarely used. Luftwaffe squadrons had gone north for the invasion of Yugoslavia, Greece and, later, USSR. We got high-level raids by Italian bombers in formation during the day, but not every day.

We had occasional strafing early in the morning and once in the evening, as far as I can remember. Nuisance raiders came over, in ones and twos, most nights. Hurricanes went up and sometimes made contact—occasionally a bomber was shot down. I remember watching such an event. Machine gun fire was exchanged, however it didn't seem a very dangerous war that was being fought.

The return of the first Luftwaffe units in late September 1941 changed all that and by November we were aware we were facing a quite different enemy. From then until the following April we were subjected to continuous air attacks by day and night. Very few ships got through with supplies and rationing was enforced.

The Luftwaffe used Junkers 88s and Messerschmitt 109s.

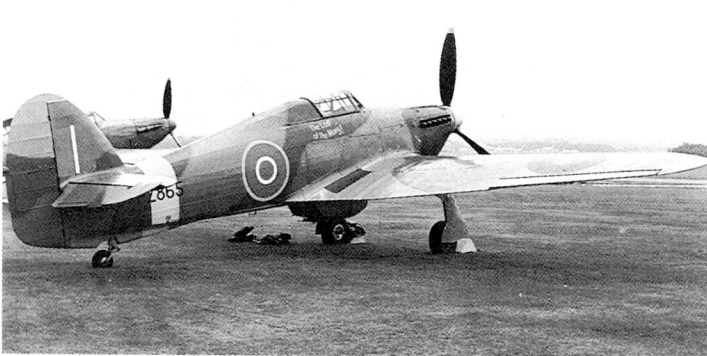

Hurricane.

An Italian Macchi fighter. Captured and in use by the USAAF.

The Italians used Macchi 200s and 201s and, later, 202s. They were mainly seen on strafing raids.

The defences were Hurricanes—inadequate against the 109s—Bofors and army and naval heavy ack-ack, or flak as it became termed after the German name. The ground defences were very good. There were batteries of searchlights located all across the island. Hostile aircraft were over or near the island most of the hours of daylight on some days. It put a great strain on the fighter pilots and ground crews. The pilots were not food rationed. The ground crews were.

Luqa, which had operated Blenheims and Wellingtons, with a few Beauforts, in the slack period, had only a few Wellingtons left by 1942. It was used as a staging post—aircraft passed through at night. Dispersal areas were extended. An area known as Safi strip extended one and a half miles towards Kalafra. A small village (Kirkop) was located right on the edge and was bombed, as were the villages of Luqa and Mosta (near Taqale).

We did not have fighter squadrons until mid-1942. A system was devised so that everyone knew when an attack on the airfield was imminent. A warning siren was sounded and a red flag was flown on the control tower. If there was an attack elsewhere on the island, or there were only a few aircraft indicated on the radar, a yellow flag was flown.

At this time Messerschmitts would come down low over Luqa, strafe the airfield and attack any aircraft landing. This became much more common later. A friend of mine was working inside a Wellington one day during a yellow alert. He heard the unmistakable sound of a Messerschmitt cannon, machine-gun and Bofors fire. He and the others got out of that aircraft very quickly and rapidly distanced themselves from any parked aircraft.

At about this time some ground crew were attempting, without success, to put out the fire in a burning Wellington loaded for a trip. It had been set on fire by incendiary bullets from 109s. Having decided it was hopeless, they were running for their lives. They

had left it a bit late and one of them was about 200 yards away when it exploded. He was blown through the air like a leaf for about 100 yards and sustained bruises and abrasions all over his body. A piece of the bomb casing had sliced open the skin above his left ear—a quarter of an inch to the right and he would have been killed instantly—one inch to the right and he would have been decapitated. He wore a bandage round his shaven head for weeks.

There was a red-headed Irish armourer who constantly told us what bastards the English had been to the Irish for hundreds of years. We used to ask him why he had joined up. He had a kneecap shattered one night when a Wellington returned from bombing the Sicilian airfields and followed the normal practice of opening its bomb doors after parking and before switching off engines. On this occasion an anti-personnel bomb, which had hung up over the target, dropped out and exploded. This was a not unknown event, but he was convinced it was another cowardly attack on the Irish.

I was on duty one night when an aircraft from Gibraltar, with a green crew, came in to land and was overshooting. It got a red Verey light. There was a Ju-88 cruising round, I could hear it quite plainly. It dropped a string of bombs which I also heard quite plainly. I threw myself to the ground. The first landed about 200 yards away, on the cross runway. I could hear the bomb splinters whizzing overhead and falling all round but I wasn't hit. The rest of the string fell in the dispersal area but didn't do much damage.

Ironically, the pilot landed on the runway that had been hit and went straight into the bomb crater for which he was indirectly responsible. He got a roasting from the wing commander in charge of operations whose caravan and slit trench were only one or two hundred yards from me.

On another night it was very quiet except for the distant throb of Junkers' engines carried on the breeze from the west. Despite the distance, I distinctly heard a voice shout 'Fire!' It came from the heavy naval gun battery near Taqali, about four miles away. Immediately afterwards came the sound of the guns and the barrage. I didn't see the first gunflash but I saw plenty afterwards.

Many of us watched, fascinated, one night as a Ju-88 was brought down entirely by searchlights. First one light held and then another, despite twisting, turning, diving and climbing they never lost it and passed it from one to another. It was soon caught in a cone of lights. I do not remember a shot being fired. The pilot must have become blinded and disoriented as, after perhaps ten minutes, he crashed on the outskirts of the village of Safi, not far from the airfield.

Later that morning some of us went to inspect the wreckage. It was scattered over a comparatively small area. The bombs had been jettisoned and there had been no fire. Bits and pieces of the crew were still to be seen. The people who lived in the surrounding villages—in fact in the whole island—became used to these episodes, although there was still weeping and wailing and gnashing of teeth.

One of the fighter squadrons, a new one being brought in, lost two of its commanders. I saw one of them shot down in flames. I saw a Hurricane sitting on the tail of a Ju-88 at about 500 feet with a Messerschmitt sitting on his tail. The Ju-88 went down into the sea but so did the Hurricane.

Luqa airfield had army personnel in the area for the period until the invasion of Sicily (July 1943). At the time it was expected that Malta would be invaded. There were a battalion each of the Devon and the Royal West Kent Regiments. At one time the Royal Irish Fusiliers were there. One of the latter said to me at one time that he, a member of the RIF and me, a member of the RAF, made up the RIF-RAF.

There was always the fear of invasion but it was also said that it cost so much to keep Malta supplied, at near starvation though it was, that the Germans were happy to keep it that way. They certainly had very heavy casualties in Crete and were perhaps not too keen to try it on again. Bofors ammunition was rationed and some retained in case of an aerial invasion.

I think it was about that time that we had to carry gasmasks to our workplaces and rifles had to be kept ready for use at any time. Many of us collected whatever ammunition we could find after the armourers had been at work. I had explosive ammunition as well as armour-piercing and normal rounds.

This apprehension was the reason that three fast minelaying cruisers were brought into play to bring in urgently needed medical supplies, ammunition, gun barrels, powdered milk, etc.—even petrol sometimes.

They were the HMS *Scotsman*, *Welshman* and *Manxman*. Between them they kept the supply up for about a year. They came into harbour and were unloaded straight away. They were never in harbour for more than one day.

One was attacked off Tunisia by about twelve Ju-88s but was not hit. There were two near misses which nearly lifted the stern out of the water and made her heel right over. They didn't think she would come back up, but she did.

450 Squadron was a Kittyhawk fighter squadron which served in the desert campaigns, Syria, Sicily and Italy. It had Australian ground staff. One of these was Roy Denny. His experience resembles a Cook's tour—a very interesting and extensive one:

I was with No. 2 Stores Depot when we were equipping 450 Squadron and, for my sins, I was posted to join the squadron at Williamtown Air Base in February 1941. The next thing I knew was that we left Williamtown and sailed out on the *Queen Elizabeth*.

They certainly had us fooled as to where we were going. We were issued with both blue and khaki clothing. We had more gear than you could poke a stick at.

We ended up in the Middle East, landing at Port Tewfik. The 'burst' [rumour] was that we were going to Greece and would be equipped with Hurricanes. Later on the rumour was that the Hurricanes were on the wharf in Athens, which didn't help much as the RAF was coming out of Greece quite rapidly at that time.

As we were without aircraft we were sent to the permanent RAF station at Abu Sueir (near Ismalia) which was the station for the Suez canal zone. There we were integrated into the station staff and did maintenance work. Suddenly we were reassembled and put on a train to Palestine to take part in the Syrian campaign.

We still had no aircraft but 260 Squadron had aircraft without ground staff. They had been torpedoed while on their way out to the Middle East and were delayed. So the 450 ground crew took their place.

260 Squadron operated V12 Hurricanes from Damascus and Rayak. The latter had been the base for the French aircraft we were now fighting—the Vichy French.

I was responsible for security, rations, spares, etc. I ordered them, chased up the deliveries and 'found' them where necessary.

The RAF operated a system of air stores parks. A wing or group of squadrons was always accompanied by an air stores park which moved along with it and was responsible for supplying its requirements. Bombs and ammunition were brought up by a separate unit which had a repair and salvage unit (RSU) attached to it.

A lot of cannibal maintenance went on. This wasn't so bad in the Syrian campaign because we had captured a lot of equipment—clothing, etc., mostly French, but we had to dispose of it when the military government people came round.

We took over the stores of the whole French station at Rayak. They had a lot of Marylands and some earlier English aircraft. I got the job of going through the whole store with my boys and packing up all the Maryland stores, with anything else we considered useful, and shipping it back to Egypt where they were operating Marylands.

After that we went to the defence of Haifa. The fuel line came from Iraq and down to Haifa where then oil was pumped into ships. At one stage we operated an airfield just below the Golan Heights. Then the 260 Squadron ground staff turned up. We were still without aircraft so they sent us to Burg el Arab. It was the end of the railway line to the desert at that time and was the site of the famous fort used in the film of Beau Geste. We used to get four-day dust storms there.

We became part of the Desert Operation—General Ritchie's push. The main strike aeroplane, believe it or not, was the Blenheim.

We had an RSU at Burg el Arab—we were called an Advanced RSU. The poor buggers would go out and do what they had to do, but if they got shot up they could land at Burg el Arab instead of going back to their base. Our job was to repair anything that landed.

We had an English Engineer Officer. We used to call him 'Puss in Boots' because he always wore desert boots. We went out on a salvage trip one day and picked up a Hurricane which was in pretty bad repair. He decided

he'd restore it. We finally got it serviceable and put it in our aircraft returns to HQ. At first they said, 'No, it's been written off.' But they finally agreed to bring it back on strength, so we had to get it air-tested. A Kiwi, who had never flown a Hurricane, volunteered to air-test it and stood it on its nose when he tried to land. So we had to fix it again. It did go back into action.

Burg el Arab is in the desert, not far from Alexandria. We got our supplies by truck from the main air force equipment store which was in caves at Helwan.

It was funny in a way. The British and German forces flowed backwards and forwards—there were quite a few pushes Wavell's, Ritchie's and Montgomery's—and we still sat there quietly in the desert repairing aircraft at the RSU.

We were relieved by the South African Air Force and went down the canal to a landing strip near Ismailia. Here the squadron got Kittyhawks and trained for about two months before we moved up to the desert. The CO and the pilots were Australian.

I think it must have been about this time, before we went up to Sidi Haneish, we were on a landing strip by the Fayum road to Cairo. We had five Chev trucks—and they were hopeless. The dust used to get into the engines and they were as weak as water. The truck bodies were all right but the engines were unreliable. So we 'made an approach' to the Helwan Transport Pool. The bloke there said, 'There are five new engines in crates over there and I won't be looking at them for twenty-four hours.'

So we whipped the engines out of the crates and took them to the strip where we had everybody who could wield a spanner stripping down the trucks and fitting the new engines. We put the old engines in the crates and replaced them within our allotted twenty-four hour span.

Kittyhawk in flight. The attachment beneath the fuselage is a long range fuel tank.

We went to Sidi Haneish, just east of Mersa Matruh, and formed our base there. We weren't what you'd call a mobile squadron with our own transport but we had two flights stationed at Gambut, which was below the Sollum Pass, at the entrance to Knightsbridge and on to Tobruk.

We operated from Gambut for three or four months, perhaps six. My job at the time was to get our supplies from the base up to Gambut. You could drive up and back in a day. If you got away at dawn there was every possibility that you'd be back before night. This was very desirable because we were right near the front line and the fun started at night.

We used to be strafed at night, not so much on the road as at the drome. We were sitting quietly in the mess one night when an aircraft came over. We knew it was a German from the unsynchronised motors. He came in low and everyone thought we were going to be bombed. However, he landed on the drome and gave himself up. It was a Heinkel. He landed in the dark without a flare path and didn't knock any tents over.

We took him over to the mess, gave him a meal and a few beers and took him into custody. We had a Heinkel triple-one for a while—until they came and took it away from us.

At that time we had 112 Squadron RAF, 250 Squadron, 3 Squadron and, later, 5 Squadron, SAF. This was the original 239 Wing, the original Desert Air Force.

However, when the Jerries broke through after the fall of Tobruk, just before Alamein, we had to get out of there very smartly.

I had a truck and my stores unit was a big trailer which I wasn't going to leave behind. Nor did I want to leave behind some canteen supplies which we had— a truck load of beer.

Bluey Matthews, who was in charge of the party, and I felt we just couldn't leave the beer; so we drank it and postponed our departure to first light the next morning. It was a big night that night. The Germans had broken through but we seemed to take it calmly.

We finally got on our way with our truck and the stores trailer. Now the truck was a Stateside one and had high wooden sides around the tray. We had with us two of the boys from Gambut and they had to ride in the back of the truck. They were, to put it mildly, jittery and kept complaining that they wouldn't be able to get out in a hurry if we were attacked.

Just before the breakthrough we had been told to take the VHF sets out of the Kittyhawks and scrap them. But we stacked them against the sides of the truck. They came to about a foot from the top of the sides. We sat our two passengers on the sets. They weren't too comfortable but they could get off in a hurry if necessary. So we saved the sets and stopped the whingeing. Everyone was satisfied.

We headed off down the road and got to a place called El Daba, which was a big supply dump. There was a big road block there to try and keep the way clear for the Australian 9th Division to get to the front.

We got to the block and the MPs tried to stop us. The two blokes in the back were as nervy as hell and screaming, 'Don't stop.' So I said to my driver, 'Give it a go, Ossie.' and we charged through without knocking anything over.

It was getting late as we waddled up the road with our trailer so we pulled off into a wadi, got under the truck and tried to sleep.

We were overlooking El Daba and saw the Focke-Wulf's come over, drop flares and wipe the place out. We were sitting on the top of the Hill of Jesus (Tel-el-Eisa), later to feature in the Battle of Alamein.

Next morning we got the old truck going again. When we got on the road to Cairo we saw a truck coming towards us. It was Terry Mullins, the transport pool chief. He was out looking for us. It was funny because we were coming down the road and he couldn't turn round. He had to keep going and caught up with us a half-hour later.

We finally got to Landing Ground 96 (Amiriya) which was just outside Alexandria.

Then everything started to point towards Alamein. Our squadron's job there was to escort the Bostons and Marylands that were softening up the Alamein area. The flying time from our drome to the Alamein area was about twenty minutes. The planes used to do about four trips a day.

After that I dropped out of the picture for a while because I contracted a desert condition that we would call hepatitis today. As soon as I was right I arranged to go straight back to the squadron and avoid the toughening-up school which was usually attended before rejoining a unit.

When I got back I was sent to a recreation place just outside Alex. It was great because it was next to the 8th Army Cookery School. I was there for a fortnight and lived like a king.

When I came back Alamein was on. Things had taken a new perspective as far as the squadron was concerned. The first twelve months of the war that we were in was operated from Shepheard's Hotel in Cairo, without a doubt.

Then Montgomery came along. Cunningham took over the fleet and Coningham, the Australian, took over the RAF. The whole thing changed.

Montgomery didn't stay at Shepheard's; he was up at the front line, living in a caravan. You'd drive along the road and he'd pull you over and have a yarn. They reckon he had more hats than you could count. If the Australians came along he'd have a slouch hat on. He was a great morale boost. This was very important because we never knew what was going on before he came. Once the three new fellows took over we used to have meetings at which the Intelligence Officer would get up and tell us what was happening. He didn't have to give great detail, such as 'The Germans are only 20 kilometres away,' but just give a general picture. Then we felt we were in touch and it helped enormously.

When Montgomery came out they formed the 8th Army and 239 Wing (ground to air close support). From there on we were a fully mobile squadron. Although we only had eight transport drivers we had just on sixty vehicles. Every unit, in the flights, in the HQ, in the stores, etc., had its own vehicles and its own drivers. I couldn't drive when I went over there. I now went through driving lessons and drove all the way up to Tunisia and up Italy too.

We still had our dirty big trailer. We went up the road and 'acquired' a Fordson six-wheeled chassis from a South African unit. We tossed the trailer away and put a caravan body on the chassis. I don't think we ever put that chassis on our books.

So we were very mobile. It is on record that, at one stage, the squadron operated from ten landing strips in a fourteen day period.

How was it done? The whole pattern of our staffing changed. The ground crew were organised into three flights, A, B and intermediate. But A and B flights were no longer assigned only to A or B flight aircraft. We adopted a leap-frog technique.

While operating at one drome, for example, all aircraft might be serviced by B flight ground crew. Meanwhile, A flight ground crew would be at our next airstrip waiting for the army to clear it. As soon as it was clear they would move in and prepare to service aircraft. When set up, all aircraft would use that strip while B flight ground crew prepared to move on to the new strip.

The intermediate flight, of which I was the proud owner, had five trucks which carried the cookhouse and all the things that matter, like bedding and tents.

As soon as the aircraft took off on their final flight from the first strip, we left for the forward strip and would be established there for the returning aircraft and B flight ground crew when they arrived.

The next move would be the same but B flight ground crew would set up first at the newly cleared strip and A flight would catch up.

We could move out in twenty minutes. Instead of the usual four or five-man tent we were issued with bivouac tents. We dug a slit trench for each tent and pitched the tent over it. Thus there was a floor to sleep on and if anything unpleasant came over you could put your tin hat on and roll into the trench.

Instead of using kitbags I got a lot of canvas and made sleeping bags—a bit bigger than normal. Then you could put all your clothes in them, roll them up like a big swag, and away we'd go.

The canvas came from old tents. Tents were always being brutalised. If you wanted a new tent you were supposed to hand in your old one first; but it didn't always happen that way.

It was the same with clothing. If you wanted three or four shirts you'd go to the stores and hand in three or four that were torn and be given three or four new ones. The stores people would put your old ones on a heap at the back of the store. We had our little black boy round there and he'd knock off four from the heap, so we were always ahead.

In this way we went right through the Desert Campaign as far as Cape Bon.

In this period we did one trip which was particularly interesting. There was a story going round that the Germans were going to come up through the Chad Valley, which is south of the Sahara Desert, and attack the 8th Army from the rear. So we were sent down to a place on the edge of the Sahara Desert which is a second Mecca of the Arab world. The main thing was to keep the Arabs on side—and that's all we did. The German threat did not eventuate so we maintained friendly contact with the Arabs and had some good hunting in the Chad Valley.

Finally the Germans in North Africa capitulated and left a lot of equipment behind.

However, you had to be careful. While driving along a road you would meet a German truck coming towards you. It would stop and a team of German soldiers would pile out. Generally they would have a German lieutenant in charge who would indicate that they were looking for a POW camp. We'd take them to a camp, tip them out and pinch the truck. We got a couple of Opel trucks that way.

Then we moved into Tunisia. We had an Engineering Officer who spoke fluent French. He said, 'Come with me. I know where we can get some good billets.' He was dead right. We were put up by a Frenchman who represented Johnny Walker whisky and we had a great time.

I remember we went to Carthage to have a surf near there. We had our host's young (twenty-year-old) daughters with us. When we got to the beach it was covered with thousands of naked soldiers. We beat a hasty retreat to another beach.

The squadron's next stop was a place called Zuara, down towards Benghazi. We were re-formed there but couldn't go on leave as Sicily was coming up.

Three of us went away and did a Combined Operations Course in Cairo. To return to Base we were to fly in a New Zealand Lockheed Hudson. The plane before us, a mail plane, took off and went straight into the side of a hill. our NZ pilot came out and said, 'OK chaps, you're next.' Seeing the looks of horror on our faces he added, 'You won't catch me doing a low circuit like that—once we're up in the air we're off.'

The emphasis was on waterproofing equipment. We used old gas capes, which we cut up, and waterproofed all the stores and packed them in ammunition cases. We waterproofed the trucks—e.g., we extended the exhaust pipes above water level.

We left Zuara soon after this and went to Malta where we operated from Luqa for a month or so. Things were hotting up and it was very demanding. Our Kittyhawks were so dispersed it took them nearly half an hour to get to the strip. We had to put a couple of blokes on their wings to guide the pilots because of the restricted visibility from the aircraft on the ground.

It was quite an experience. The airfield was very busy and, among the traffic were shot-up Liberators returning from raids on the oilfields in Poland. If there was a prang there was no question of salvage. The end of the runway was a sheer cliff and the wreck was just pushed over the edge into the sea.

We were billeted in a rather elaborate place—it was a lunatic asylum.

Our next stop was Sicily. We landed on Cape Pachino which is right on the tip. The old skipper of our landing craft said, 'Don't worry, I'll put you ashore without getting your feet wet. I've been tramping up and down the Mediterranean all my life.' He ran the landing ship up on the shore and we just drove the trucks off and walked off. We nearly got trampled to death by the POWs trying to get on. Apart from that we had no trouble at all.

We were a reasonable distance from the fighting so we camped in an olive grove. We had a landing strip just up the road. A few aircraft came in and we started operating straight away.

We carried arms—a small American light rifle. We also had revolvers which we took on the ration run.

We had had a variety of people guarding our dromes. The RAF Regiment at one stage and the Gurkhas as we came up through Africa—aerodrome defence was

their rest period. For one three-month period we had a South Pacific group from Noumea. When the squadron was reorganised to become mobile we were given army guards again—often Gurkhas.

As the war moved on we moved up to Agnone, just near Catania. This was one of those places where there was only one landing strip and it went out into the sea. There were hills all round it. We operated quite a lot from there. We couldn't camp on the aerodrome so we camped on the hills and went down to the drome every day. It was lucky we did so because the Germans really belted the drome at night.

The time came for us to move on again. They were having trouble at Anzio and wanted air support. A British Commando unit took the aerodrome at Brindisi, put a perimeter round it and we flew in.

We went over in a DC-3. I'll never forget that. We were so low, if you had put your foot out of the side door it would have got wet. During the flight an Italian Macchi came up towards us—we thought we'd had it. However, he waggled his wings and flew away—much to our surprise. We didn't know Italy had capitulated.

Flying time from the drome at Brindisi to Anzio was like a trip into town—six kilometres. The Kittyhawks did four or five trips a day. They played a big part in the taking of the bridgehead.

We then moved on to a similar situation—Monte Cassino. We were on an aerodrome just near the Mount and there again the flying time was about twenty minutes or so. The units servicing us had the game by the throat.

Then we went on to Rome. We were very fortunate there. The Germans declared Rome an open city but nobody believed them. So they moved us into Dadahnga, which is just outside Rome.

Fred McKay took the Australians in to see the Pope. He put on an audience for 3 Squadron and ourselves. Quite a number of our squadron went and he asked us questions. Our three padres, Catholic, Anglican and Presbyterian, went with us. Incidentally, the three of them were later picked up by the military government for knocking off materials from some villages to provide a chapel—but nothing came of it.

After Rome we moved on up the coast to mainly farmland areas. Then the Jerries looked like breaking through on the Adriatic side. There was a wild flap, and we shifted the whole wing, in one day, from the west to the east (380-odd kilometres) to a drome near Vasto.

We operated from there and then moved to a drome just north of Florence. Then I left them to go home via Cairo and 459 Squadron.

I came home via Bombay, where the Yanks were to pick us up and take us to Australia. We waited three months in that transit camp with blokes from Burma (an Australian and an American crew), AIF, POWs (120), and about 100 Australian aircrew, all going home.

In the end, I had a good trip back.

As it turned out, I sailed from Australia on Good Friday 1941 and got off the train at Central in Sydney on Good Friday 1945.

CEYLON

The leave provisions for ground staff were invariably poor. Home leave was virtually out of the question on Far Eastern stations.

David Corthorn tells how this policy affected some of the ground staff with whom he had to deal:

I was Orderly Officer of the Day at my station in Ceylon. I had to inspect the airmens' quarters—they were mainly English ground staff who wanted to get back to their wives.

I went into a large hut and they were all sitting up in the rafters, putting on a great act of pretending to be troppo. If you weren't in the know, you'd accept they were mad.

I saw the CO and he said, 'Don't worry about them. They put this act on every time a new officer comes in. They think he might be a bit naive'.

SOUTH -WEST PACIFIC

Papua New Guinea and the Islands

Ground staff in this area of the war did not have the same 'luxury' of travel as did many of their counterparts in other theatres. Added to this were the problems which were particular to their area. There always seemed to be a shortage of necessary equipment and this led to long hours of work under the most adverse of conditions.

For example, the fitters in the Catalina squadrons often had to do all the work on their planes—or 'boats'—on the water as slipway facilities were simply not available.

Norm Cromack tells of some of his experiences while serving in New Guinea during the dark early days, later in Queensland and, finally, in New South Wales:

There were two squadrons of Cats at Port Moresby, Nos 11 and 20. These squadrons stayed together from the time they were formed in 1940 until the end of June 1944. They were in Moresby before the Japs came into the war, patrolling the islands and Rabaul. At one stage in January/February 1942 there were just three Catalinas

and two Hudsons to keep the Japs out of Australia.

In those early days it was the practice to paint the hull below the waterline with linseed oil to help preserve it. But it was found that, when aircraft went to Rabaul, pumice stuck to the sticky coating and this didn't make for smooth landings.

One aircraft broke its back up there. A sergeant fitter—a permanent air force bloke—cut down a tree and trimmed out a log which he put down the centre of the aircraft. He braced it all up and the aircraft was flown from Rabaul to Rathmines.

I left Townsville on Friday 13 February 1942 and arrived at Moresby three or four days later. Initially the Japs didn't bomb us at the harbour but concentrated on the Seven Mile strip. That changed later.

It's a lonely feeling when your working on a Cat in the water and there's nowhere to go if trouble comes. You're too far out to swim in—certainly with overalls on—and swimming's too slow in an air raid.

I remember one raid early in the piece when they came in a straight line for us and the wharf. The ack-ack put up a barrage and the Japs turned away. I don't think they really meant to attack us. But the next raid was different. It was 28 February and I had not been there long. I was off sick. The RAAF put five or six aircraft up over the other side of the bay. The Jap fighters strafed the dispersed Cats. Two sank and the rest caught fire. We managed to get one ashore and used what we could for spares.

We were hoping for Kittyhawks to give cover. We called them Tomorrowhawks because they were always coming tomorrow.

In another raid the Japs dropped a bomb right near the tail of a Cat. The water came up and broke off half the tail. We took the tail off the burnt Cat, fitted it on the damaged one and a crew flew it back to Rathmines.

We had two fitters who used to do this work. Our ground staff had it all over the Americans with whom I worked for a short time in Moresby. Our ground staff categories (Group I Armourers, Airframe Fitters, Electricians and Instrument Makers) were all specialists in their jobs. The Americans had crew chiefs in charge of teams of non-specialists. They were amazed at what we could do and we knew far more about the Cat motors than they did.

I was in Moresby at the time of the Battle of the Coral Sea (5-8 May 1942). We knew something was going to happen as our Cats had spotted the ships. All air force personnel were evacuated to a radar station eight or nine miles out of Moresby. We were told that, if the worst happened, we were to go west to the Fly River where we would be picked up.

We carried rifles but were not skilled in using them and only had a few bullets. As it happened, the Japs didn't bomb Moresby after all but they did strafe our drome. I remember this well because we didn't have slit trenches. The noise was horrendous.

During this period the RAAF was stretched to the limit for aircrew. The aircrew we did have would come in the morning and have to go out again that night. We didn't have crews as such; we only had pilots. There were no navigators in the early days—until they found that they needed them. They carried two pilots but one had to do the navigation. We had one chap, a squadron leader, who was an expert navigator. His first pilot was a sergeant.

I was there when the Japs sank the *Macdhui* (18 June 1942) after first crippling her steering. Our marine section was the only rescue effort available and they took the wounded off.

A corvette went aground on the reef just outside Moresby and the Japs tried to get it. They came over at 18 000 feet—the lowest I'd ever seen them. The ack-ack fired six shots and brought three down; I think they were using radar. The first shot blew the leader right out of the sky. He was the bomb aimer for the group and when he was shot down the rest just dropped their bombs in the sea nowhere near the corvette.

The ack-ack were militia men and very good. The Japs tried in vain to blast them out.

The Japs attacked our own ship one day. It was the air force supply ship *Wanaka*. They chased it all round Moresby harbour but didn't get it. One of the men on board told me later that when it was at the wharf and the Japs were seen coming over the skipper had only two lines holding the ship and a man with an axe at each line.

He got the engines going at full power and said, 'Cut them.' She came out from the wharf like a speed-boat. Incidentally, the people who had been unloading her were still on board. The skipper lay on his back on the bridge with a pair of binoculars. As the bombs left the planes he'd order 'starboard' or 'port.'

There was a little Dutch ship in the harbour one day when we were warned that a hundred Jap aircraft were coming. The sirens went and twenty-six bombers came over Moresby (some went over Horn Island) with about twenty fighters as escort. They pattern bombed the harbour. We could hear the bombs coming down and exploding.

When the raid was over, there was the Dutch ship unhit, but covered with mud and muck. The skipper complained to me, 'My beautiful white uniform is all covered in mud and dirt.' Apparently he wasn't concerned about his ship.

The Cats used to carry two mines. They were fitted with radio altimeters and were given a reference point at which to start their mining operations. They would fly over the reference point at a fixed course, speed and height. After a fixed time interval they would drop one mine, continue on for another fixed time and drop the other.

Mining tied up Japanese men and boats sweeping for them and, because they could not be sure they had got them all, disrupted their supplies and their effort.

The Americans had Airacobras which carried a dirty big 37 mm cannon. In one attack on Moresby we could hear a chap attacking the Jap bombers. They jettisoned their bombs and did more damage than if they had been allowed to bomb normally.

I first struck the American negroes up there. This conversation took place when one of our drivers was giving a negro a lift—he was driving pretty fast.

'What size shoes do you take?'

'Why?'

'Because they push awful hard on that accelerator.'

A fitter IIE was responsible for the maintenance of the engines of all aircraft. You were bound by rules and regulations and had to abide by them. Inspections were daily, preflight, 40 hours, 80 hours, 120 hours and 240 hours, depending on the motor, and a major inspection after 700 hours in the case of the Catalina.

We used to do our 80 hours inspections on the squadrons. A maintenance unit at Bowen did our 240 hours inspections. When we were at Rathmines we could do our 240 hours inspections on the squadrons because, unlike Cairns, we had a slipway.

We had a marine section which took us out to the moored planes. It also refuelled the Cats. Mines and bombs were taken out in scows. We often worked on planes with mines or bombs aboard.

We were out in the sun seven days a week. Sunburn was common.

We went out on the water every morning to do the daily inspection. Even if the aircraft had been inspected yesterday morning it had to be inspected again.

The routine was: lift off the cowling, see there were no oil leaks, make sure nothing had come loose, make sure all the locking wires were intact. Everything on the motor had to be locked so that it wouldn't come undone with the vibration. Some of the parts were locked together with locking wire, some had castellated nuts. There was a drain on the Catalina petrol filters so that they could be checked for water in the fuel.

The plane had always to be checked before it could be taken out so that a check had to be squeezed in if it arrived back late at night and was required next morning.

Airacobra.

The aircrew would plug up any holes below the waterline with rubber plugs—only those below the waterline, the rest didn't matter.

The aircraft had a kit of tools on board because you had to have special tools. You could just about take a Pratt & Whitney engine apart with a 7/16 BAE spanner but there were some parts you couldn't do anything with unless you had a special spanner.

We used to change the spark plugs every 80 hours. A spanner about eighteen inches long was used to tighten them up but we used one about two and a half feet long to break the seal to get them out. If we broke the spark plug and left part of it in we had something like fourteen nuts to undo. This was difficult when the aircraft was in the water.

You have to be very careful with a radial motor. There is a master connecting rod and every other connecting rod is fitted to the master. If the master rod is taken out you have to be very careful to not drop it in the water and lose everything. We used to tie them

on to the aircraft and be very careful. We often sent them down to Bowen where they had a slipway.

The Catalinas were modified to give them a longer range (about 4000 miles). All the armour plating and the self-sealing tanks were removed.

It is interesting that there were self-sealing hoses in the motors. If they got a little crack in them they would self seal and cut off the petrol to the motor. They gave us no end of trouble. In the end we took them out. When they swelled we ripped them out and put ordinary hoses in.

At one time we had a lot of trouble with one motor. It had just been reconditioned back in Australia but in normal flight it would suddenly suffer a partial loss of power. We looked at everything but couldn't find the cause. We sent it back to Australia but when it came back it was just the same.

Aircraft engines have two separate ignition systems and if one fails the motor still runs but with reduced power, just as happened with this motor. We had taken one distributor off one day and an electrician was just looking at it when someone pressed the booster button. This activated a series of high-voltage pulses in the ignition system to produce sparks at the spark plugs before the magnetos became active when starting up. These pulses also produced sparks in the distributor. The electrician suddenly realised he could see the sparks in the distributor. There was a hairline crack that had escaped everyone's notice. It was pure luck that the fault was found.

I left Moresby in September 1942. The Japs were just at Kokoda at the time. I know because I worked with the Yanks at the time at the Twelve Mile strip. We used to sleep with knives and arms under our pillows. We were the furthest airstrip out. We were twelve miles out and the Japs were only about twenty or thirty miles out.

We were salvaging aircraft and we lingered on. The food was good but we were sick of baked beans. We used to swap the baked beans, which the Americans loved, for more palatable food. I enjoyed the pancakes and maple syrup.

At the beginning of 1942 Port Moresby was very ill-prepared for any attack. In Chapter 4 we have shown how few resources were available to provide for the air defence of Moresby.

The problem was compounded by the lack of airfields. There was no civilian labour force to construct them and the army engineers hadn't the resources.

Thus the air force was called on to construct its own airfields. This they started at Port Moresby.

Syd Kildea, an LAC electrician, was in No 5 Mobile Work Squadron at Moresby at the time. His story traces the early history of the unit:

I started at Wagga in 1942 with No. 1 Mobile Works Squadron Special Works Force. The mobile squadron was formed because the army couldn't cope with building dromes during the push in the Pacific.

We thought we were going to Darwin so I read up on 'We of the Never Never'. Instead we went to Moresby on 25 July 1942. Later, as the Air Board had one mobile works squadron in Darwin and one in Moresby, signals were getting mixed up so we were renamed No. 5 ACS.

The trip from Melbourne to Moresby was something to remember. We were in the *Abiel Foster*, a Liberty ship of 10 000 tons. On the way from Townsville to Moresby we got stuck on the reef. We were sitting there for sixteen hours—we were sitting ducks, the reef light was shining on us all night. The next day a little boat came alongside, took our anchors and dropped them alongside

5. Aircraft Construction Squadron en route to Port Moresby.

us. It then tried to pull us off, finally succeeding at three o'clock in the afternoon. We'd been there since eleven o'clock the previous night. We later learned that the ship that pulled us off (the *Kaloola*) was torpedoed a few hours later (4/5 August 1942). The ship behind us was sunk by a Japanese sub on the day we got to Moresby.

Some of the Liberty ships broke in halves on the reef. Our officers had decided that if we had been there another night they would have loaded us into the ship's boats and landed us on Cape York.

Townsville had been bombed quite recently (25 July 1942—the day we left Melbourne), Moresby on 29 July and Mossman (near Cairns) was bombed on 31 July. Another ship was sunk around the NSW border on the same day that Townsville was bombed. We came into Townsville on 2 August 1942. We were sailing in dangerous waters.

On arrival at Moresby we disembarked and marched in the heat to the RAAF barracks. Our ship got stuck on the reef in Moresby harbour and had to be unloaded by barge. They got it off the next day and we spent the next few days completing the unloading.

A couple of days later I came down with chicken-pox with a temperature of 104 and they put me in hospital for three weeks. I didn't want to report sick because there were some reporting sick just to get out of work and I didn't want that reputation.

When I got out the place was a hive of activity. We built Ward's Field in Moresby. It was originally a dummy strip with dummy planes on it to serve as a decoy. Seven Mile drome (which became Jackson Field) was the main drome then. Ward's was begun on 7 August 1942. It was an all weather asphalt strip.

I looked after the alternators and ran the power. We worked a twenty-four hour day. We had to have our

Syd Kildea's photos from the front. *Above:* Port Moresby Hospital where he spent two weeks. *Below:* Tractor and overturned carry-all, Ward's Field contruction. *Bottom:* Unexploded bomb at Port Moresby, 1942.

mobile alternators down on the strip to allow work to proceed and we had a 25 KVA alternator in the workshop. In the early days we only had small earth-moving equipment but the situation built up as time went on.

Our men came largely from civil construction companies and local councils. The head of RAAF Construction Units was Wing Commander Dale. He was put in charge of all advanced field construction units at the direction of General Kenney, US Air Force.

Early in Moresby the cry was always, 'When are we going home?' Some of the chaps reckoned, as the Japs were coming over the Owen Stanleys, that they had their bags packed. They reckoned they'd walk home. It was all bad news then. It was a great boost to morale when the Japs were stopped at Milne Bay and on the Kokoda Trail. There were some Kittyhawk pilots in the RAAF hospital at this time. They had been shot down over Milne Bay and suffered severe sunburn.

We didn't have any poles to hang our cables on. We were told we ought to run them on the trees and that sort of caper. We did finally get poles and we scrounged some conduit from a hangar in Moresby.

They called us the Flying Shovels because that is what we started with (picks and shovels). Service life taught me to scrounge around for equipment—from the Yanks, from bombed-out houses—from anywhere in fact.

We (the workshop) stayed in Moresby for some eighteen months. Later we moved on to Goodenough and then, as the war progressed, things moved on quickly.

Our CO was an architect and all our officers were engineers. The CO claimed to be a water diviner. He found water, and lots of it, right under our camp.

We had sixty raids while we were there. The planes were in revetments and the bombs mainly hit the runways. They came over with up to about twenty-five bombers when there was half-moon. We used slit trenches for protection and had very few casualties

It was funny that in Sydney there was a brownout but in Moresby you could see the trucks coming back from the pictures with their headlights full on and visible for miles.

Food was basic—baked beans 'pregnant goldfish', bully beef, herrings and tomato sauce.

At the end of August 1942 we heard on the radio that nine Japanese ships had come into Milne Bay. It looked like things were getting worse but at midday the next day we heard that the ships had come to take off troops, not to land them.

On 13 September Beaufighters landed at Ward's and played a big part in the Kokoda/Buna campaign.

I had one go at biscuit bombing. It was in November 1942. I'd given an army chap a lift to Ward's Field and

Insignia of the "Flying Shovels", No.5 A.C.S. (Formerly 5 Mobile Works Squadron).

heard there that we'd retaken Kokoda. They were going to drop supplies so I decided I'd go along too—unofficially.

I was down the back of a DC-3 pushing boxes out the door when one of them got off to the left. As we pushed it hit the door which was held back by a rope. The door came off its hinges and wrapped itself round the tailplane. I was worried that it would get jammed in the controls. I knew of cases where the tailplanes had been knocked off and the planes had crashed.

The co-pilot came back and said, 'We'll get back.' We took half our load back. We first tried to land at Kokoda but they waved us away. Possibly they were still clearing mines from the strip, although they were landing there and using the drome the next day.

I was in Moresby at the time of the Battle of the Bismarck Sea. I've never seen so many planes in the air. They flew from Townsville to Moresby the night before so we knew something was on but we didn't know what.

There were Fortresses, Liberators and whatever else they could lay there hands on. We heard later that the Beaufighters and Bostons from 22 and 30 Squadrons strafed at sea-level while the Liberators and Fortresses came in over the top. It stopped the Japs from reinforcing Lae.

We kept working for months after the Kittyhawks came in to extend the Jackson strip for bombers. By the time we left, in March 1943, it was a formidable drome. It was a big job done in a hurry under very trying conditions.

Goodenough Island is situated a short distance north of the Papuan coast. It is small (about sixteen miles in diameter) but quite mountainous. In late October 1942 the 2/12 AIF Battalion landed there and established control. The Japanese had been there since their failure at Milne Bay.

Two RAAF mobile works squadrons landed on the island to prepare an airstrip. Syd Kildea was with one of them:

They decided, just before Christmas 1942, to send us to Goodenough Island. Our CO said we'd been picked to go because we were crackerjacks. The name stuck and we became known as the 'Crackerjacks.'

We landed on Goodenough in April 1943 and built Vivigani Airfield. At that time we were further north

than any other force in our area (and possibly than any in the Pacific). There a few Jap stragglers from the Bismark Sea Battle but they were up in the mountains. They tried to do some damage but were unsuccessful.

We were armed with .303 rifles and the army used to go looking for the Japs—it was a small island.

We built a wharf at Bola Bola. We had to send a team to Fergusson Island to get the timber.

Besides supplying power for our work we supplied it for lighting all the mess huts and some tents but we did not supply power for cooking.

Our men were working long shifts and boredom was a problem when they stopped. The army provided a 35 mm projector so we built a projection booth. We got old films like *Captains Courageous*.

We put the same film on seven times a week and our fellows would go to every show. I reckon that if the actors had made a mistake our people would have corrected them.

The films the army sent us were ancient. I remember one in particular, *Lucky Partners* with Ginger Rogers and Ronald Colman. We had it for five days. Then the ship didn't come in with new films so we had it for another five days but the boys still came every night to see it.

The Yanks used to get the latest films but had 16 mm equipment. We used to borrow their equipment and films to break the monotony.

The Yanks had plenty of equipment but they used to blow up their projectors—they weren't well versed in electrics. Their projector was a long way from the alternator, which meant the current had to come through a lot of wire, and then they used a light wire to feed the projector. To add to the problem they were using 110 volts.

When they turned the projector on the picture would fade so they upped the alternator speed. When

Goodenough Island—camp at airfield construction site.

Movie Theatre, Goodenough Island.

they changed a reel the lamp went off, the load on the alternator fell, the voltage went up to about 160 volts and 'Poof!' would go the projector. We made sure we had heavy cable all the way to our machine. We became friendly with the American who ran their projector and helped him with his electrical problems.

The USO[1] shows used to come to the American camp so my mate and I went over to see if the artists would come over to our camp. We spoke to 'Little' Jack Liddle, an American singer/composer—he wrote 'A Shanty in Old Shanty Town'—and to Ray Bolger who was in *The Wizard of Oz* and they agreed to come over and give us a show.

The boards on our stage were rough timber planks and Ray Bolger was coming to dance. We said, 'We can't ask him to dance on such a rough surface.' So we went to the Barracks Officer and asked if we could borrow a couple of sheets of masonite which, like everything else up there was as scarce as hens' teeth. He said we could have it but it had to be returned that night.

They made all the difference to the stage. Ray Bolger put on a terrific performance. I can't recall taking the sheets back.

There was a negro outfit there. They worked all day as truck drivers unloading on the wharf. One of them had played with Cab Callaway's band. He came over, bringing his bass, with a group at 9 p.m. one night to do a show for us. It was really marvellous.

After Goodenough we came back to Melbourne.

In March 1944 I went from Sydney to Milne Bay and Lae, which had only just been recaptured. But as soon as I got to Lae I was posted back to Wagga and then to Narromine.

Jim McSharry tells of an experience in another area of the South-West Pacific—Morotai Island between New Guinea and Sulawesi:

The Americans were the major task forces in the area, and had a policy of taking just enough of an island to

1. United Service Organisation: a private body dedicated to the welfare and entertainment of armed forces personnel.

establish the facilities they wanted. They saw no point in taking the whole of a big island—expending lives to acquire real estate they didn't need. They took enough for their requirements, and then established a guarded perimeter around the area left to the Japs.

This was the situation at Morotai, where there were three airstrips, a big port, and a big army area, set up in about 20 per cent of the whole island.

The result was that we never had the luxury of having a substantial dispersal area. The general practice was to line the aircraft up, wingtip to wingtip, along the taxi ways. There were some Jap bombers in the area, but the powers that be reckoned this was a fair bet and saved taking more of the island.

By the end of February 1944, we'd been operating from Morotai for some time. We were wanted for an attack on a Japanese-held island in the Celebes—just up from Morotai—probably less than an hour's flight.

We were to take off at first light, so, during the afternoon, all eighteen Bostons were bombed up with a volatile explosive called RDX. All the machine-gun belts were filled and the petrol tanks topped up. We were ready for a dawn take off—the whole squadron.

We had some air force guards on the squadron. One of these was a very young man—probably eighteen—who had desperately wanted to join the air force. He wanted to be a pilot, but by that time they weren't really training any more pilots. It was getting late in the war and there were lots of pilots coming back from Europe. He did a guard course and volunteered to go into the war zone.

Because of his interest in aircraft, he used to haunt the airstrip when not on duty. He would stand on the wings while they started the Bostons. Because the pilots were some distance away, the sergeant fitters were allowed to taxi the aircraft which had to be moved for servicing. One of these sergeants had befriended the boy and had allowed him to sit in the cockpit with him while he, the sergeant, manipulated the controls and taxied the aircraft.

The lad was on guard at the 22 Squadron strip at Morotai, on the 4 p.m. to midnight stretch. A single Japanese light bomber came over at about 9.30/10 p.m. It attracted a lot of searchlights and a lot of flak and let go a string of bombs. Unfortunately, one of these went through the middle of our line of Bostons and set one on fire. That aircraft blew up and set the two adjacent ones on fire—it looked like going right along the line.

The guards had taken shelter and the corporal of the guard tried to get through to the camp to warn them of the danger to the aircraft.

The lad jumped the fence, ran to the nearest Boston, kicked away the chocks, found a foothold, got into the cockpit, pushed the primer levers and opened the throttle to just turn the engine over. The first one started and started the battery charging and let him start the other engine. He taxied the aircraft out of the fire and shut it down.

He then jumped down, ran back to the other end and did the same thing with a second Boston. Amazingly, he did it a third time. By some miracle, he saved three Bostons, loaded with RDX, from this enormous conflagration, with bombs and machine-guns going off. The CO stopped him from doing any more.

The squadron was reduced from eighteen aircraft to three.

AUSTRALIA

Service at home in Australia, though it had its obvious compensations, was not necessarily a 'cushy' job. Doubtless, life in an orderly room, with many clerical and administrative chores, could be rather quiet. However, the orderly room staff were often witness or privy to developments which were exciting and which had far-reaching effects.

Andy Emmerson, trying to speed up his entry to aircrew, became an orderly room staff member:
I was ground staff, office orderly, to start with. I was eventually remustered to clerk general, and then became an orderly room sergeant. I then remustered to aircrew.

I was one of the first recruits to go into Amberley, in Queensland. The camp was brand new—they didn't even have any beds. They gave us palliasses, which we filled with straw, and we slept on those.

After my recruit training, I went to Archerfield and joined 23 Squadron. It started off with two flights of Wirraways and one flight of Hudson bombers. Later, it changed over to Airacobras, and eventually ended up as a Liberator squadron.

Archerfield, with its galvanised-iron huts, was like an outpost. When war was declared by the Americans, we dispersed all our aircraft. We saw the American squadrons come in, and the formation of 75 and 76 Squadrons. The first Wirraway squadron to go to New Guinea left from there.

I was there when the first American squadron came in. The bombers arrived from America just after we'd had some intense tropical downpours and the aerodrome was just mud. They were short of fuel. Our CO was up in a Wirraway, trying to direct them to Amberley, where there was a sealed runway. The first decided to land, and went straight through a fence and knocked out a house at the bottom of the runway. The second headed towards Amberley and crash-landed.

We later saw the Douglas dive-bombers come in. The Americans had all their pilots mixed up—bomber pilots were being trained to fly dive-bombers. their performance wasn't too good but, at that time, there was a great degree of urgency because of what was going on at the top [the Japanese advance on Port Moresby].

One day, at Archerfield, we had all these shiny new Wirraways lined up. the Engineering Officer came in and told the CO they were all serviceable. there was to be a training bombing run that day.

At that time, we also had an Elementary Flying Training School operating Tiger Moths on the drome. One of them was taking off, swung 90 degrees, and crashed into the Wirraways. Dixie Chapman was our CO (he later went to 3 Squadron). He raced out, firing red Verey cartridges into the air. In the end, he had all the incoming Tiger Moths circling like a bees around a hive, not knowing what to do. It was quite a performance.

The first Hudsons went to New Guinea from Archerfield. I remember one pilot came back to Archerfield with his Hudson riddled with bullets down both sides. He had been attacked by Zeros and his only way out was to fly from one cloud to another, and then head for home. He had then to fly south for repair and re-equipment. The Hudson pilots were mainly ex-Qantas pilots.

In 23 Squadron, at Archerfield, we had the Jackson brothers. Each commanded a flight of Wirraways. One of them was shot down and arrived back after some weeks, having walked through the jungle.

Keith Hampshire was CO at Archerfield during the early stages, when the Americans were coming in. He was doing evaluations of the Lightnings, Kittyhawks and various other planes. He used to put them through aerobatics over the drome, and I used to type up his notes. He used to refer to planes as 'gun platforms' or 'bomb platforms' and, as far as the pilots were concerned, they were just the jockeys who had to deliver the plane to the target.

I was still at Archerfield when my transfer to aircrew came through.

Although Australia was a centre for production, planning, training, organisation and supply in the war against Japan there was also an extensive operational role, with all the attendant hazards.

Norm Cromack returned with his squadron to Cairns in September 1942. He gives some details of his service life there until the beginning of July 1944:
Bombing up was quite a task. The armourers used to lift them up to the bomb racks. They generally used a hoist as they were often 500-pound bombs. The mines were anything up to eleven hundred pounds.

When we were at Cairns the wharf labourers went on strike because they wanted danger money for unloading bombs. An armourer said, 'I'll fix these blokes up.' He took a tip truck out to the dump and brought back a load of bombs and tipped them on the wharf beside them. The labourers disappeared quite quickly. When they came back he said, 'That's how dangerous they are.'

Of course they could have been dangerous if they had been 'weeping.'

Sometimes the aircraft took daisycutters. These were bombs with a rod eighteen to twenty-four inches long protruding from the nose. The idea was that the rod would hit the ground and set off the detonator with the bomb still above the ground. It could clear an area of thirty to forty feet. I once saw one fall off a plane before take-off. You've never seen an aircraft move so quickly to get airborne.

We had to work over the water at Cairns. You were supposed to put a lanyard on your tools but you never did. There was a funny experience concerning this.

I came back from leave and found one of my team had no eyebrows left. It appeared that he was, as usual, cleaning the back of the motor with petrol to get rid of the oil and grime. While he was doing this someone pushed the booster button. There was a burst of flame and he dived into the water to save himself. He had carefully placed his tools on the catwalk but in the excitement his rescuers knocked them all into the water.

We could take some leave but were always on call. I was out at a friend's place in Cairns one night. A truck pulled up outside at ten o'clock and the driver said, 'Norm. You're wanted. There's an aircraft which has to go in the morning and the pitch control isn't working.'

Working at night on the water wasn't much fun. Our lighting was the aircraft lights and an auxiliary power unit which we used to start the motors. The batteries on the Cats were very small and had to be conserved. Sometimes we worked by torches. You couldn't do a major job that way.

It was twelve midnight when we finished. We were cranky with this pilot and his last-minute demands so the WO in charge said, 'We'll fix this bloke.' He telephoned the officers' mess and said, 'Come down and air test your aircraft.' But the pilot said he'd take our word for it. It was a pity really because we'd have loved to see him sitting on the water in the early hours of the morning belting the throttles open to maximum power.

Rathmines in New South Wales was very much an operational base for Catalina squadrons. Coastal surveillance (which included anti-submarine patrols) was carried out from there.

Norm Cromack was there from July 1944 to the end of hostilities. He tells his story:
When we came from Cairns to Rathmines we had to improvise as we had very little equipment. There was quite a lot of 'borrowing' of equipment between units. In an attempt to stop this each unit painted its portable gear a distinctive colour. We painted ours green. When we 'borrowed' other equipment we quickly painted it green to make it obvious that it belonged to us.

We had trouble organising platforms for our aircraft inspections. However, we found some old shed doors and used the iron from these to build our platforms. It was most embarrassing when a detective came down to investigate where some missing iron doors had gone. He leaned on our platforms while he questioned us but didn't catch on.

6

THE TACTICAL AIR FORCE IN EUROPE

On 6 June 1944, the greatest invasion in history took place on the French beaches in Normandy. The forces were made up of the Second British Army (which included the Canadian 3rd Division) and the First American Army. There were five beach attacks and landings, the Americans at Utah and Omaha, and the British at Gold, Juno and Sword. Nearly 7000 vessels and 130 000 men made up the invading force.

The first stage of preparation for a tactical air force, to support the landing forces in the invasion of Europe, began in June 1943. All of No. 2 Group, Bomber Command, was transferred to Fighter Command on 1 June. Both Fighters and bombers were now under the same control. 464 Squadron RAAF (Venturas) was one which went over to Fighter Command at this time.

Advance Into Belgium

After landing and consolidating their position, the Allies broke out of the beachhead as Caen fell, on 9 July 1944. Montgomery's Second Army moved up through Northern France and Belgium into Holland. It was in support of this move that Ron Goode served, on Tempest fighters, in 2 TAF

The Germans deployed very effective flak defences with their ground troops, and these were a very significant problem to aircraft attempting to give close support to the Allied forces. Ron describes the opposition he faced:
I joined 80 Squadron (Tempests) 2 TAF in Holland. 80 Squadron was a very old RAF squadron which had been in existence from 1917. At this time, it was part of 122 Wing, which was comprised of five squadrons. One of them was a New Zealand squadron. There were supposed to be twenty-seven pilots to a squadron, but we were losing planes so fast that we could only average twenty pilots per squadron at any one time. The flak was so bad that, in a twelve-day period in February 1945, we lost

100 aircraft and thirty-three pilots.

The Germans had ringed Europe with flak and, as we forced them back, their anti-aircraft guns went with them. We'd encounter heavy flak practically every time we went out. You'd look ahead and think, 'Christ! It's going to hit me!' So you'd look left, and it was there as well. You'd look right and it was there, look behind and it was the same. It was 40 mm flak—'red tennis balls' we used to call them. We didn't fly above 8000 feet—four thousand more likely—and were very vulnerable to it.

We didn't lose so many to fighters because the Germans didn't attack us unless they had twenty plus. We flew in squadrons of eight—two flights of four would fly in formation. Our squadrons would go out at ten-minute intervals.

Ron had good reason to be thankful for the New Zealand squadron in his Wing:
One day, we heard a New Zealand voice saying, 'F...ing 190s.' [FW-190s] so we knew there was a heap of 190s between us and base, looking for us, and the NZ squadron beyond them. So we prepared to drop our long-range tanks, when, all of a sudden, eight planes went down in flames, and it was all over.

The NZ squadron had come up behind them while they were concentrating on getting us.

Attacks on Long Range Rocket Launching Sites

Bomber attacks on V-1 flying bomb sites were referred to in Volume I. These attacks on Britain ceased on 1 September 1944, but were replaced by attacks with the V-2 long-range rocket. First launched on 8 September 1944, some 1190 rockets were eventually fired. About 500 fell on London.

Second Tactical Air Force was called on to attack launching sites, mainly round the Hague, from where most rockets were launched.

Ron Goode, flying Tempests, and Andy Emmerson, flying Spitfires, both attacked rocket sites. They tell their stories.

Ron Goode:

I trained on Typhoons at a Typhoon/Tempest OTU and subsequently flew the Tempest, which was the fastest piston-engined fighter in any air force.

We'd be flying over Germany on a beautiful summer's day with no enemy activity in sight when, all of a sudden, we couldn't see anything for flak coming up out of the trees. We couldn't see any guns—it was just a wood from which a gun had opened up.

It could be a V-2 site, so we'd give it a squirt—unless we had a specific target to go to, in which case we'd ignore it.

Andy Emmerson:

I was posted to operations with the Australian Spitfire squadron (453) in late 1944. We were part of the 2nd Tactical Air Force.

We flew from an aerodrome in Norfolk, called Matlaske. The strip didn't have solid runways and was not always serviceable. I had my first prang there—I got caught in the mud and the Spit went over on its nose. As soon as it hit the ground it was just sucked into the earth. There wasn't much left of the aircraft. When rain made the drome unserviceable, we had to fly out of Swannington and had a long drive there every day.

We were mainly concerned with dive-bombing V-2 sites. These were very difficult to locate. They were transportable and were put up against hospitals and various other unlikely places. We depended on the Underground to give us their location. Invariably, when we went out in the morning, the Germans would be starting off their operations.

The German ack-ack was very accurate and we lost a lot of pilots in our attacks. the German gun crews were particularly accurate on our run in. As we approached the Dutch coast on our run in from England we oscillated, diving and pulling away, to elude the guns.

We were always vulnerable because we were going in at about 12 000 feet and diving to about 2000 feet. Invariably, as we pulled out, the German gunners concentrated on one plane and followed it down. So, when you came out of the blackout, you looked around to see if there were any tracers and if it was your turn in the barrel, as we used to say.

We did have a number of people who were shot down by the flak. The Spitfire was the greatest defensive fighter in the world. However, as an attacking fighter, it was very vulnerable to flak. It had an in-line engine and the slightest sliver of flak could cause it to lose coolant—and that was the finish.

If we got shot up, we would do a belly landing. The Dutch Underground were fantastic. I think that, out of eleven pilots who were shot up and belly landed during one period, not one was captured by the Germans.

They had young boys who would come and guide the evaders into the woods. One of our flight commanders was actually billeted in a house where there were German officers. The Dutch would transport evaders by motorbike. They put them in the sidecar and sat on them.

At that time, the Allies were coming through, and there were strips where you could land. I got into trouble one day and my number two said, 'There's an aerodrome over there. He's giving you a green light.' But my response was, 'There's no way we're going in there. That's Dunkirk and it's a German-occupied drome.'

We were losing a lot of aircraft and pilots through forced landings, so Command said, 'Go out to sea and we'll pick you up.' Although this sounded illogical, we did it.

This leads me to the story about 'Swampy' Marsh. He was Air Sea Rescue Officer on our squadron. He trained us on dinghies, air sea rescue and so on. Eventually, he was shot down and he baled out over the sea.

We had to look for him and we soon realised how difficult it was to locate a small dinghy in amongst the waves. He was off the Dutch coast, but not far enough out for a safe rescue by boat. We located him and our Air Sea Rescue sent out a Walrus. It got down and landed, but was knocked out by the German guns. So the crew got into their dinghy and floated around.

Later, the RAF sent out a Catalina, which landed but had one of its engines knocked out. So it started to taxi away, picked up the Walrus crew, and escaped.

'Swampy' was still there that night. He'd done all his air sea rescue drill, aerial up, radio beeping away so an aircraft could home on him.

Four of us went over next morning. We had been given an indication that he'd drifted in fairly close to the Dutch coast. We flew up and down for about ten to fifteen minutes before I finally located him. It was purely by chance I saw something out of the corner of my left eye. He'd put his hand in the water and splashed some up—so, there was 'Swampy', in his dinghy. I climbed and radioed in, but HQ considered he was too close in to try to rescue. So they agreed to a truce with the Germans, who sent out a German Red Cross boat. We went over later in the afternoon and saw the German boat, and there was a pilot waving vigorously. But it turned out that it wasn't 'Swampy'—it was an American Thunderbolt Pilot.

'Swampy' drifted ashore next morning, but after a couple of days in the water, he could hardly walk. The Germans didn't treat him too kindly either.

During this truce period, when we flew over to monitor the rescue, we saw four midget submarines coming out on the surface. We didn't do anything about them because of the truce. The next day, when we flew out, the sea was aflame and there was smoke to about 30 000 feet from a torpedoed Allied tanker. I often question whether they took advantage of the truce.

I had a couple of near misses with landings. I landed the plane on its belly once. I was just off the coast of England, and because the coast was mined, I was trying to get down in a small paddock a little inland.

The field was not very long, so I put my wheels up and flaps down and dived off my height very quickly. When I levelled out, I had just enough speed to get in. But, at the other end of the paddock there were a number of trees, so I had to come down abruptly. The aircraft completely disintegrated and I ended up with just the cockpit. The wings fell off.

A farmer was ploughing a field nearby and he came running over. I felt like a cigarette and offered him one. His hands were shaking—mine weren't.

I put this down to training. We were drilled to do a forced landing, and I just carried out the landing without even thinking about it. It was just second nature.

The Tactical Role

The role of the 2nd Tactical Air Force led it into sorties of varied character, with targets of many types.

The requirement of the Supreme Command was to disrupt communications and transport by attacking railways, roads, troops, tanks etc. The enemy must be starved of reinforcements and supplies.

Ron Goode, Jim Dransfield and Andy Emmerson develop this.

Ron Goode tells of the varied targets:
We attacked trucks, tanks and troop movements. We came in from a couple of thousand feet, at 500 mph with four 20 mm cannon.

The day we did the tanks there were whole heaps of them. The fellow who led us went down and we followed in a straight line. I was Arse-end Charlie and I said 'Christ! I'm going to get it.' I was as scared as those Germans were—and nobody could be more scared than that. We stopped them, anyway. They weren't going anywhere when we left—but we didn't hang round to review the position.

I nearly shot myself down once. It was right at the end of the war, on one of the last trips we did, a job at Dosfold. It was late in the afternoon. We had a fellow from Chile with us, an ex PRU pilot, but he had no idea of flying formation. He used to get lost every time because he'd shoot off on his own.

The three of us attacked a train and it stopped. We looked back, and there were animals running everywhere in the field. On closer observation, they turned out to be Germans, so we gave them a squirt on the way through, but found that there were now only two of us.

The weather was lousy, but we saw another train coming our way. We attacked it, but I got a bit carried away. I was after the engine. With a train, you had an engine, then a flak car, some carriages, another flak car in the middle, some more carriages and a flak car at the end. The flak cars would give you some swish. I attacked the engine and blew the boiler off it. But I got too carried away and just pulled up in time. I could see gun fire near me. It was my own shells ricocheting off the engine. I nearly shot myself down. It was easily done.

I thought it was time we left, as it was getting pretty late. Just then, we saw another train. Although there were all these balloons with their steel wires ready for us, I said to my partner 'We'll give it a go, but we'd better hurry.' So we did, but he got hit and had a bloody great hole in his wing. The flak took off the pitot head, so his instruments weren't any good. I asked him, 'Can you hold it up?' He said, 'I can hardly hold it at all.' So I said, 'OK Follow me,' and we went through a hole in the clouds and tried to get in touch with base.

I could hear base but couldn't transmit, but we always steered the same course (270 deg.) home. Once we got past the Rhine, it didn't matter. I eventually went down through the clouds, and we got in at about a quarter to nine. It was just on dusk and bloody dark—they had given us up.

My partner was holding on by his pants and I said, 'I'll lead you in, and then I'll go round again.' I did a very quick circuit and came in.

The CO was waiting—he was a prize prick. He said, 'A bit of flak, eh?' He could kiss my arse. We got very full on brandy.

Ron Goode had another frightening experience:
I lost my No. 1 the next day—he went down. We rode shotgun on a group of Typhoons (Rocket)—they went 300 miles into Germany. They had never been that far. We could do it because we had ninety gallon drop tanks each side—as they did—but they were frightened all the way.

I've never seen a more abortive trip in all my life— they panicked. 'How much further? We haven't got that much juice,' and so on. Eventually they attacked—they only hit one thing.

Typhoon Fighter.

We were fed up. The leader of our four planes went looking for action. I was flying No. 2 and I could see a train sitting there with steam coming out of it. There was a flak tower at the back, and it looked funny to me. My leader, who was an experienced squadron leader, and I, dived to attack but as I pulled up I saw he was on fire. He hit the ground just as his parachute opened.

The other two planes in the flight hadn't bothered to go down—which was just as well, as it was a flak trap. When we looked back, we could see the track was busted. The train was not going anywhere—the Germans just kept a fire in the engine for deception.

There I was, on my own in the middle of Germany—the other two buggers had pissed off—thank you very much—and I was feeling a bit lonely. I steered 270 deg., between the Hun and the sun, and eventually I saw two spots quite a distance away. These two buggers came up alongside. My R/T had gone useless—I hadn't realised that I could hear everything but that I couldn't transmit.

I heard radar predicted flak—bang, bang, bang,— metal on metal—and my aircraft went up in the air. I looked down and there were two more bangs. I put the throttle through the gate and the last two flew over me. I had certainly heard metal on metal, but there was not a mark on my aircraft and I never did find out who had been hit. We didn't know where we were, but once across the Rhine it was virtually safe.

All of a sudden, I saw the twin towers near where we were stationed—and I wheeled her in.

Jim Dransfield, at this time, was also engaged in harassing the enemy transport as well as attacking military convoys, tanks and troops:

We were flying Mosquitoes, in effect, as a back-up to the army. When it moved further towards Germany, we moved up to Brussels (Melsbreek). At night, our squadron, and other members of the Intruder Group, patrolled a particular section of the air space over Holland or Germany to attack any evidence of rail or road transport bringing supplies or men to the front.

There were two attacking armies—Montgomery in the north, and Patton in the south. Our support was usually directed to Montgomery's army.

Attacks were usually initiated by the sighting of vehicle headlamps. The Germans would try to draw us into traps by, for example, opening the firebox of a train which was made up solely of anti-aircraft guns. We usually used parachute flares—we'd open the bomb doors and drop the flares, then descend beneath them and attack with machine-guns—always remembering there was a danger limit when the navigator would tap you on the shoulder.

This may sound a bit casual, but part of our training was to make sure we didn't kill ourselves. The navigator was schooled to concentrate on the altimeter as it unwound, and belt his pilot around the shoulders if ever the plane descended into a dangerous area—such as, within 1000 feet of the highest point known in the area.

We had four Hispano machine-guns (20 mm cannon) and four .303 machine-guns in the nose of the aircraft. The parachute flares were for use in machine-gun attacks. If we carried flares, we had no bombs—the flares filled the bomb bay. If there was a bombing raid, the bomb bay was filled with four 500-pound bombs.

More often than not, losses were due to not following the right procedures. On my first trip, the navigator didn't watch the altimeter—he was more concerned with what I was doing. When I suddenly realised we were in danger, I pulled out. But I must have had a wing down and I flipped on my back. I don't know how I got of that, I think it was divine intervention.

We operated at night, mostly, but some of the senior pilots did individual sorties in the daytime—for recreation.

We attacked whatever we saw. For example, on the night of the crossing of the Rhine, our orders were to put our noses down and attack the German barracks and billets to keep them out of bed. We were at the Nijmegen end.

My main danger was from the Americans. At each end of the combined front there was free air passage. Any Allied plane was to be given automatic freedom to enter or leave Germany. Flak was to be shut down.

But, on several occasions, while in the southern sector, near the Americans, I got bursts of flak. The only way to stop it was to give them a few bursts from the machine-guns in return.

Ron Goode also attacked motor convoys:
I had a stopped motor truck in my sights one day. The driver had dived to the side of the road, but one bloke stayed, leaning on the side of the radiator, looking up at me. I blew up the truck. What a death wish he had. I just couldn't believe it. The Tempest weighed five tons and had four 20 mm cannons—that's a lot of firepower.

Talking about luck, I chased a convoy along a road which went into a village. I got a bit carried away, I was still firing after I'd hit a truck and, as I pulled up, I almost went into a three or four-storey building. I almost brought myself down. Then the whole bloody building exploded—a terrific explosion. It must have been full of ammunition. It frightened shit out of me.

I'll bet the Germans said, 'What a wonderful source of information they must have had. How did they get to know about it?' It was sheer luck.

Ron Goode tells of what might be euphemistically be called 'mistaken identity':
Every six weeks or so, we got a week's leave in England. On one such occasion, I came back on a Dakota with another Tempest pilot.

He flew that same afternoon and was shot down, when on the way home, by an American Thunderbolt which came up from behind him. The standard of aircraft recognition in the US Air Force was bloody awful. Although a Tempest was not unlike a FW-190, it was a very much bigger aircraft.

Anyway, when he was shot down, his mate swooped round and shot down the American. Every time a pilot fired a cannon, an automatic camera took a photograph to record what was hit. So, when this chap landed, he said to the armourer 'Destroy that film.'

Ron Goode was lucky to escape from a crash while operating:
Our CO was a real panic merchant. In fact, they used to call our squadron 'The Panic Squadron'. He finished up as the Commander of the Wing. Mind you, he had a good record—it was just that he was so keen. We only flew twelve days out of twenty-eight, but we were out at the crack of dawn every day. We lived about five miles from the drome. We'd bolt down our breakfast and go out in a truck to sit out there in this bloody fog and rain. All the other squadrons were back at the monastery where we were based, taking it easy, but Panic Squadron was out there in case it cleared up and we could get away.

Everything in the area was flattened. There were millions of hand-made bricks used in our runway, which was sixty yards wide and a mile long. There were two runways, which the Germans had put down using forced labour. Our mob had repaired any bomb damage to them, but there were dozens of bomb craters around them.

I was coming in to land one day, when, all of a sudden, there were flames shooting out from under the wing. I gave it full right rudder and full brake. The left wheel had come off and I had landed, on one wheel and one oleo leg, at 120 miles per hour. The oleo leg grinding on the bricks was spitting out sparks everywhere.

I had 1000 hours up at that stage, if you can't fly then, you never will, but I had only about thirteen hours up on bloody Tempests. You weren't supposed to fly operationally without twenty hours up on the type, but being experienced, I didn't worry about it. If I'd made a crook landing on this occasion I'd have been in real bother.

Attacks on Gestapo Headquarters

The Danes, whose country had been occupied by the Germans since 1940, carried out considerable sabotage to embarrass their enemy. However, the Gestapo collected a great deal of information and their files posed a threat to members of the Danish Resistance. With the advance of the Allies, it became more and more likely that the Germans would seek retribution and move in on this organisation.

The Danes asked for help. Three specific raids were carried out to provide this. Jim Dransfield describes how his squadron participated in the attack, by eighteen Mosquitoes and twenty-four Mustangs, on 21 March 1945:
The senior pilots were, usually, older and more experienced men than I. Most were family men and were

determined to get home to their families. They were really more value to operations than were the young flibbertigibbets like me. Whenever there was a specific mission, other than a patrol, these men would be rested from operations to give them time to study the mission and, in particular, mock-ups of their target. One such mission, for example, had been the breaching of the prison wall at Amiens, which allowed imprisoned members of the Maquis to escape execution. While I was on the squadron, doing the night-time patrols, the squadron was briefed to attack the Shell building in Copenhagen.

After waiting a considerable time for suitable weather, they took off for the sortie and flew at contour level (basically, zero feet), only altering course to avoid power lines known to be in the area.

The task was for each to put a bomb through the windows of the building. They did this very successfully. The only Danish casualties was a group of little children in the playground of a convent school. They, most unfortunately, were killed when one pilot, a Canadian from our squadron and on his third tour, crashed and killed himself and his navigator, as well as the children [eighty-eight children and twenty-four adults were killed].

I went to Copenhagen after the peace and was welcomed with open arms. I was, after the war, associated in business with a Dane who had escaped detection by the Gestapo because of the destruction of records during this raid.

The squadron was involved in a similar raid on the Gestapo Headquarters in the Hague in Holland.

Passive Resistance

The German occupation forces were not popular with the people of the occupied countries. We have seen how Resistance forces helped Allied airmen who landed in Holland, and how the Danish and French Resistance movements were helped by intruder action. But there were other, less covert ways of frustrating the Germans. Jim Dransfield tells of one such way:

In May 1945 we were billeted on a Luftwaffe establishment which had been built by the forced labour of Belgian workers. The Germans thought they had put it all over the Belgians, but the Belgians got square.

They made sure that all the sewerage pipes ran uphill. Every day there was a large pool of shit which could not get up the hill. It had to be siphoned out regularly by tankers, taken away and dumped. It was dumped all over the place.

In those days, the Belgians still used pots under the bed. In the day time they'd empty the night's accumulation over their little domestic vegetable gardens. We were warned to not eat anything from domestic gardens because of the danger of disease. It wasn't too difficult to dispose of the sewage by random dumping, but there was a hell of a fuss one day, and a great traffic of Germans to nearby farms where, it was alleged, two tankers were surreptitiously delivering black market petrol.

The Occupation

Andy Emmerson relates:

We were told on VE Day that we were not going home but going to Germany with the army of occupation. When questions were asked in the Australian parliament the answer was that we were volunteers for the duration and twelve months thereafter. There was quite a deterioration in morale when we got to Berlin.

It was not a nice place to be anyway. It had just been captured.

The first place we got to see was Belsen Concentration Camp. We were stationed at a place called Fassburg. One of the flight commanders had a jeep and we went to Belsen. If anybody thought the Germans didn't know about it they're wrong. It was only half a mile down the road from a German NCO school when we were there.

All the inmates of Belsen had moved out into various compounds according to nationality, but the gas ovens were still there. They took us down into the cellar where they had the Beast and Beastess of Belsen and their colleagues. You'd never see such an ugly mob of cutthroats. They were the people who had perpetrated this. I could never forget Belsen.

In Berlin we met with the Russians. They had raped widely in Berlin. We were stationed at Gatow. There was nowhere to go. If you took off in a Spitfire you had to make a rate-two turn to stay in the British Sector.

In Berlin we couldn't go out with a pair of wings on. We heard of bomber crews which had allegedly been lynched.

We did enjoy the company of the Gordon Highlanders who had come through from the desert. It was a very much decorated unit.

While stationed at Fassburg I saw a lot of Germany. I went to Lubeck where, they said, all the churches had been bombed. On examination it seemed they had been used to store bombs, ammunition etc. We weren't very welcome but we were flying the flag.

Our country club had been Goering's country residence with 400 acres of deer forest. We used to go out shooting hares and deer. We gave a considerable amount to the starving Germans.

★

C O N T R I B U T O R S

(T = Tape   W = Written memoirs   D = Diary entry)

Leo Allen
Leading Aircraftsman RAAF Armourer/Air Gunner
42926 11 Squadron RAAF
South-West Pacific Area   (T)

John Appleton
Flying Officer, RAF Wireless Operator/Mechanic/Air Gunner
578305/56466 210 & 190 Squadron RAF
Coastal Command   (T&W)

Ron Barker
Flight Lieutenant RAAF Navigator/Wireless Operator
420372 24 Squadron RAAF
South-West Pacific Area   (T)

Len Barton, US
(A US soldier medal for gallantry officially recognised)
Warrant Officer RAAF W/O Disciplinary 458 Squadron
C7841 207841
Middle East & Mediterranean Commands   (W)

Roy Baxter
Flying Officer RAAF Navigator
421879 53 Squadron
Coastal Command   (T)

Sam Birtles
Warrant Officer RAAF Navigator/Bomb Aimer
418049 454 Squadron RAAF
Middle East and Mediterranean Commands   (T)

Ernest Blowman
Sergeant RAF Station Intelligence
77 & 102 Squadrons RAF, 462 & 466 Squadrons RAAF
Bomber Command   (W)

Keith Caldwell
Flight Lieutenant RAAF Pilot
261499 Independent missions with C Class flying boats
South-West Pacific Area   (T&W)

Fred Cassidy
Flight Lieutenant RAAF Wireless Operator/Navigator
35723 30, 93, 457 Squadrons RAAF
South-West Pacific Area   (T)

Eric Cooper, AFC
Wing Commander RAAF Pilot
93 (later 260093) 7, 22 & 73 Squadrons RAAF
South-West Pacific Area   (T)

David Corthorn
Flying Officer RAAF Navigator
420452 292 Squadron RAF
Coastal Command & South-East Asia Command   (T)

Jim Cowan
Flying Officer RAAF Navigator
400109 11 Squadron RAAF
South-West Pacific Area   (T)

Bob Crawford, DFC
Flight Lieutenant RAAF Pilot
402642 12, 75 & 78 Squadrons RAAF
South-West Pacific Area   (T)

Norm Cromack
Sergeant Fitter II E
94594 12 Squadron RAAF
South-West Pacific Area   (T)

Jack Davenport, DSO, DFC & Bar, GM, MID
Wing Commander RAAF Pilot
403403 455 Squadron RAAF
Bomber & Coastal Commands   (T)

Roy Denny
Flight Sergeant Equipment Branch
32105 450 Squadron RAAF
Middle East and Mediterranean Commands   (T)

Jim Dransfield
Flight Lieutenant RAAF Pilot
422078 538 Wing RAF, which included 464 Squadron RAAF
2nd Tactical Air Force   (T)

Ted Eagleton, DFC and Bar
Squadron Leader RAAF Pilot
402727 466 Squadron RAAF
Middle East Command & Bomber Command   (T)

John Eather
Flying Officer RAAF Pilot
432 499 31 Squadron RAF
South East Asia Command   (W)

Bruce Edenborough
Flying Officer RAAF Pilot
412122 6 & 36 Squadrons RAAF
South-West Pacific Area   (T)

Andy Emmerson
Flight Lieutenant RAAF Pilot
2239 453 Squadron RAAF
South West Pacific Area & 2nd Tactical Air Force   (T)

Peter Fisken
Flight Lieutenant Observer
8013 5, 12 & 30 Squadrons RAAF
South-West Pacific Area   (T)

Ron Gardner, DFC
Flight Lieutenant RAAF Pilot
420382 680 & 681 Squadrons RAF
Middle East Command & South East Asia Command   (T)

Ron Goode
Flying Officer RAAF Pilot
412429 80 Squadron RAF
2nd Tactical Air Force   (T)

George Gray, DFC
Squadron Leader RAAF Navigator
402346 113 & 55 Squadrons RAF, 454 Squadron RAAF
Middle East & Mediterranean Commands   (T)

Bob Hartman
Flight Lieutenant RAAF Pilot
422178 Ferry pilot/ test pilot RAAF
South-West Pacific Area   (T)

John Hopley
Leading Aircraftsman RAF Instrument Repair
3010279 340/341 (French) Squadrons 6341 Services Echelon
2nd Tactical Air Force   (W)

Bill Hughes, DFC
Squadron Leader RAAF Pilot
403928 47 Squadron RAF
South-East Asia Command   (T)

Laurie Jones, DFC
Flight Lieutenant RAAF Pilot
421224 160 Squadron RAF
South-East Asia Command   (T)

Gerry Judd, DFC
Flying Officer RAAF Pilot
420957 600 Squadron RAF
Ferry, Middle East and Mediterranean Commands   (T)

Syd Kildea
Corporal Electrician RAAF
62202 5 Mobile Works Squadron RAAF
South-West Pacific Area   (T)

Ron McCathie
Squadron Leader RAAF Navigator
401325 454 Squadron RAAF
Middle East Command   (W)

Jim McSharry
Flight Lieutenant RAAF Pilot
404710 217 & 22 Squadrons RAF
Middle East Command & South-West Pacific Area   (T)

Peter Matthews
Flying Officer RAAF Navigator/Bomb Aimer
424777 454 Squadron RAAF
Middle East and Mediterranean Commands   (T)

Leslie Mills
Leading Aircraftsman RAF Electrician
1542676 1693 Flight RAF
United Kingdom and Iceland   (T)

Doug Moffat
Flight Lieutenant RAAF Navigator
412999 246 Squadron RAF 10 and 461 Squadrons RAAF
Coastal Command and Transport Command   (T)

Walter Mould
Sergeant RAF
574976 Delivery Flight Malta
Middle East & Mediterranean Commands (W)

David Nesbitt
Flying Officer RAAF Pilot
420355 5 Squadron RAF
South-East Asia Command (T)

Doug Nicol
Warrant Officer RAAF Navigator/Bomb Aimer
424446 212 & 240 Squadrons RAF
South-East Asia Command (T)

Len Parsons, DFC, MID
Flight Lieutenant RAAF Pilot
25026 100, 8 & 7 Squadrons RAAF
South-West Pacific Area (T)

Jack Robertson
Flight Lieutenant RAAF Pilot
421394 14 Squadron RAF
Middle East, Coastal & Mediterranean Commands (T)

Jim Savage
Corporal RAAF Fitter IIA/Air Gunner
15268 10 Squadron RAAF
Coastal Command (T&W)

Jack Stronach
Flying Officer RAAF Pilot
402415 117 Transport & 178 Liberator Squadrons RAF
Middle East & Mediterranean Commands (W&T)

Reg Torrington
Sub-Lieutenant RANVR Observer
Squadron RNS 853
Fleet Air Arm (T)

Spiro Tsicalas
Flying Officer RAAF Wireless Air Gunner
402625 34 & 31 Squadrons RAF
South-East Asia Command (T)

Ron Turner, DFC
Squadron Leader RAAF Navigator
413933 37 & 150 Squadrons 330/231 Wing RAF
Middle East & Mediterranean Commands (T)

AIRCRAFT

The details of performance, armament and crewing mentioned in this section are taken from the Australian War Memorial publication, *Air War against Germany and Italy 1939-1943* and the Memorial's approval for its use is gratefully acknowledged. In most cases the performance is that of the latest mark.

The details of the Airacobra, Avenger, Barracuda, Black Widow, Cant, Demon, Empire Flying Boat, Harvard, Valentia, Vengeance, Wirraway and Zero were compiled from a variety of sources, which include crews' diaries, etc., and not from any one source.

Airacobra (Bell)

A single-engined fighter, maximum speed was 360 mph. Armament consisted of one 37 mm cannon, two .5 inch and four .3 inch machine-guns. Range was 1000 miles. Crew one. (USSR, US)

Anson (Avro)

A twin-engined low-wing monoplane. Introduced into the RAF as a general reconnaissance/bomber aircraft in 1936, it was obsolescent at the outbreak of war. It was kept in operational use in the UK and Australia for some time, until replacements came off the production lines, when it was widely used as a training aircraft in the Empire Air Training Scheme. It was widely respected for its docile handling.

The Anson had a maximum speed of 178 mph and a service ceiling of 18 000 feet. It was armed with two .303 machine-guns. Maximum range, depending on load, was 600 miles. Maximum bomb load was up to 500 lbs. The crew was usually three or four. (RAF and RAAF)

Arado Ar 196 (Arado)

A singe-engined floatplane designed for reconnaissance work from ships. A low-wing monoplane with a maximum speed of 195 mph and a service ceiling of 23 000 feet, it was armed with two 20 mm cannon and two or three 7.9 mm machine-guns. It could carry about 220 lbs of bombs and had a range of about 600 miles, depending on load. The crew was two. (German)

Avenger (Grumman)

A single-engined low-wing monoplane which first entered service with the US Navy as a torpedo-bomber early in 1942. It had a maximum speed of 276 mph and a service ceiling of 22 000 feet. Armament was four .5 inch machine-guns and two .3 inch machine-guns. It could carry one torpedo or 2000 lbs of bombs. Normal range was 900 miles but could be as much as 1900 miles without bombload. The crew was three. (US and Fleet Air Arm)

Baltimore (Martin)

A twin-engined midwing monoplane built as a medium bomber and reconnaissance aircraft for France and Britain, it was only used in the Mediterranean area. It had a maximum speed of 308 mph and a service ceiling of 22 300 feet. It could carry 2000 lbs of bombs. Maximum range was 1100 miles. Crew was four. (RAAF and SAAF)

Barracuda (Fairey)

A single-engined shoulder-wing monoplane used by the Fleet Air Arm as a torpedo bomber, minelayer and reconnaissance aircraft Maximum speed was 220 mph and the ceiling was 20 000 feet. Armed with two Vickers gas-operated machine-guns on a flexible mount in the rear cockpit. It could carry six 250-lb bombs, four depth charges or a torpedo. It carried a crew of three. (Fleet Air Arm)

Battle (Fairey)

A single-engined low-wing monoplane built before the outbreak of war as a light bomber, the Battle suffered devastating losses in the Battle of France and was immediately withdrawn from operations. It was relegated to use as a training vehicle for use in training gunners and bomb aimers in the Empire Air Training Scheme. It had a maximum speed of 241 mph and a service ceiling of 24 300 feet. It was armed with two or three .303 inch machine-guns. It could carry 1000 lbs of bombs and had a maximum range of 795 miles. The crew was two or three. (RAF and RAAF, training only)

Beaufighter (Bristol)

The Beaufighter had a maximum speed of 345 mph and a service ceiling of 32 600 feet. It carried four 20 mm cannon and six .303 machine-guns. Up to 1000 lbs of bombs or a torpedo could be carried. The range was 1470 miles, this could be extended to 1810 with long range tanks. The crew was two. (RAF and RAAF)

Beaufort (Bristol)

The Beaufort was a twin-engined low-wing monoplane designed for the roles of general reconnaissance, bomber and torpedo bomber. It went into service in October 1939. Its maximum speed varied from 234 to 257 mph. It had a service ceiling of 26 000 feet. The armament varied but later versions carried nine machine-guns in the nose and two .303 guns in a power-operated dorsal turret. It could carry a torpedo or up to 1610 lbs of bombs. Nominal range was between 1070 and 1660 miles. (RAF and RAAF)

Black Widow (or XP61) (Northrop)

A twin-engine monoplane with twin tails, it was designed to be a night fighter. Maximum speed was 360 mph and service ceiling was 33 000 feet. Armed with four .5 inch machine-guns and four 20 mm cannon. Crew was three. (US)

Blenheim and Bisley (Bristol)

A twin-engined low-wing monoplane used as a long-range fighter-bomber and in reconnaissance. The Blenheim appeared in five versions during the course of the war. Bisley was the name given to the Mark V Blenheim. Bolingbroke was the name allocated to Mark IV Blenheims produced in Canada. Armament varied from version to version and depended on the task being undertaken, but was similar to that of the Beaufort. In fact, the four aircraft bore a strong family likeness. The Blenheim had a maximum speed of about 270 mph and a service ceiling of around 24 000 feet. (RAF)

Bombay (Bristol)

A twin-engined biplane designed and built well before the war as a transport aircraft, it was obsolete for use in the European theatre at the outbreak but lingered on in the Middle East. Maximum speed was 192 mph and ceiling was 21 800 feet. It could carry twenty-four troops. Range was 880 miles. (RAF)

Boston (Douglas)

A twin-engined midwing monoplane used throughout the war as night fighter, bomber and ground attack aircraft. Maximum speed was 315 mph and the service ceiling was 24 500 feet. Armament varied widely with 20 mm cannon, .5 and .303 inch machine-guns being used. It could carry a torpedo or 4000 lbs of bombs. The range could be extended to approximately 1530 miles by use of long-range tanks. The crew varied from two to four according to the role. (US, RAF, RAAF and USSR)

Buffalo (Brewster)

A single-engined monoplane fighter used by the RAF initially in Singapore and Malaya, it was a failure in the face of the Japanese Zero. Later it was used in the army cooperation role in Burma. Maximum speed was 324 mph. Armament was four .5 inch machine-guns and the service ceiling was 30 675 feet. It could carry two 100-lb bombs. (RAF)

Cant Z.506B (CANT)

Three-engined low-wing floatplane used for reconnaissance and bombing. Maximum speed 229 mph at 13 000 feet. Armament was one 12.7 mm machine-gun and three 7.7 mm guns. It could carry a torpedo or 3000 lbs of bombs. Range was 1600 miles. Crew of four to five. (Italian)

Catalina (or PBY) (Consolidated)

A twin-engined high-wing flying boat used in a wide variety of roles—particularly in anti-submarine activities. The maximum speed was 196 mph and the service ceiling 14 700 feet. Armament was two .5 and three .303 inch machine-guns. The Catalina was renowned for its long endurance and range of 3100 miles. At a cruising speed of 130 mph it could be airborne for more than twenty hours. A crew of ten was carried to cope with this. (US, RAF and RAAF)

Commando (C 46) (Curtiss)

A two-engined transport plane originally designed as a civil airliner. It had a maximum speed in excess of 250mph and a ceiling of 25 000ft. It was used in Burma as a troop carrier (30-40 troops), supply dropper and general transport. Used for flights to China. (USA)

Dakota (DC-3) (C-47A) (Douglas)

A twin-engined low-wing monoplane which was the military version of the extremely successful DC-3 civil airliner which carried 21 passengers, a development of the DC-2 which only carried 14 on introduction to Australia. A very reliable and efficient aircraft, it was used in almost every theatre of the war as an unarmed transport aircraft to carry troops and supplies, drop paratroops and tow gliders. The maximum speed was 217 mph at 10 000 feet and the service ceiling 24 000 feet. It could carry a useful load of 6700 lbs and the range was 1600 miles. Crew was three. (US, RAF and RAAF)

Demon (Hawker)

A single-engined biplane developed from a very successful family of prewar light bombers to serve Australia as a fighter and light bomber in the years before the war. It was obsolescent at the outbreak of war but was used at some OTUs for pilot training until superseded by the Wirraway. Maximum speed about 180 mph. Armament was two .303 inch machine-guns and one Lewis or Vickers GO gun. Crew was two. (RAAF)

Empire Flying Boat (Short Bros)

A four-engined high-wing flying boat designed and used solely for Qantas passenger services on the 'Kangaroo Route' between Australia and England before the war. Leased to the RAAF for the duration of the war and used for general transport purposes. (RAAF)

Flying Fortress (or B-17) (Boeing)

This four-engined bomber was the mainstay of the US bombing effort in the European theatre. A very rugged aircraft, it had a maximum speed of 300 mph and a service ceiling of 36 000 feet. It was armed with thirteen .5 inch machine-guns. Its bomb bay could accommodate 6000 pounds of bombs. It had a range of 2000 miles with a full bombload. The crew was six to eleven for bombing operations. (US and RAF)

FW-190 (Focke-Wulf)

Single-engined low-wing monoplane fighter used also in the ground attack role. Maximum speed was 435 mph at 39 000 feet. Armament was four 7.9 mm machine-guns with an option of two 20 mm cannon. Maximum range was in the vicinity of 1000 miles. Crew was one. (German)

Gladiator (Gloster)

A single-engined biplane fighter, the Gladiator was the last biplane fighter of the RAF and was obsolete at the outbreak of war. However, shortage of aircraft forced its use in the defence of Malta until production caught up in England. Maximum speed was 245 mph and its ceiling was 32 500 feet. It was armed with four .303 inch machine-guns and carried a crew of one. (RAF)

Hampden (Handley Page)

A twin-engined bomber close to the end of its useful career when war broke out. It was also used for minelaying and as a torpedo bomber. Maximum speed was 247 mph and the service ceiling was 19 000 feet. It was armed with six .303 inch machine-guns. It could carry up to 6000 lbs. of bombs and had a range of 1800 miles. The crew was four. (RAF)

Harvard (North American)

A single-engined monoplane used as a two-seat training plane by RCAF and RAF. Maximum speed was 210 mph and it had a range of about 750 miles. It carried no armament. (RCAF RAF)

Hudson (Lockheed)

The Hudson was a two-engined monoplane built as a general purpose aircraft. It was used as a transport aircraft by Coastal Command for general reconnaissance. Maximum speed was 242 mph and the service ceiling was 24 500 feet. It was usually armed with seven .303 inch machine-guns. It could carry bombs or depth charges. Maximum bomb capacity was 1400 lbs. Normal crew was four. (RAF, RAAF, RCAF, RNZAF, US, Netherlands and Chinese air forces)

Hurricane (Hawker)

A single-engined single-seat fighter, the Hurricane was the RAFs first monoplane fighter. It served in the front line throughout the war. It was the major equipment of Fighter Command during the Battle of Britain. With the steady introduction of the Spitfire and Typhoon in the European theatre, it was used right to the end of the war in the Middle and Far East. The maximum speed was 342 mph. Service ceiling was up to 41 000 feet. Armament varied from eight .303 inch machine-guns to four 20 mm cannon. As a fighter bomber it could carry four 500-lb bombs. Its range was about 500 miles but could be extended with external tanks. (RAF)

Ju-52 (Junkers)

A three-engined low-wing monoplane used as a transport, troop carrier and glider tug. Maximum speed 165 mph and service ceiling 20 000 feet. Armed with four or five 7.9 mm machine-guns and one 20 mm cannon. Could carry 4000 lbs. or 22 troops. Range was between 530 and 790 miles. Crew of three or four. (German)

Ju-86 (Junkers)

A two-engined German bomber with a cruising speed of 224 mph and a range of 1555 miles, it was armed with three 7.9 mm machine-guns and could carry 2200 pounds of bombs. A crew of four. (SAAF, German)

Ju-87 (Junkers)

Known as the Stuka, this was a single-engined monoplane built initially by Germany as a dive-bomber. However, it proved too vulnerable to fighters after the initial shocks and was developed for ground attack, mainly by night. Maximum speed was 225 mph and service ceiling was 25 000 feet. Armament was four 7.9 mm machine-guns. Maximum bomb load was 3100 lbs and range was 410 miles Crew of two. (German)

Ju-88 (Junkers)

A two-engined aircraft used by the Germans as a bomber, dive-bomber, torpedo bomber, night fighter and for reconnaissance. Maximum speed was 370 mph and the service ceiling 30 000 feet. It could carry two torpedoes or up to 6400 lbs of bombs. The armament varied with the role—a mixture of 7.9 and 13 mm machine-guns and 20 mm cannon. The range varied between 853 and 1620 miles. A crew of four was used in the bomber, three in the fighter. (German)

Kittyhawk (or P-40)(Curtiss)
A single-engined single-seat monoplane used as a fighter and fighter bomber, the Kittyhawk was one of the American P-40 line which started with the Tomahawk. It had a maximum speed of about 350 mph and a ceiling of 30 000 to 32 500 feet. It carried four to six .5 inch machine-guns and could carry up to 1000 lbs of bombs. The range was 470 miles with bombs and up to 1200 miles with long range tanks and no bombs. (RAF RAAF US)

Liberator (Consolidated)
A four-engined bomber used by the US (as the B-24) mainly in the Pacific and Middle East, it was used by the RAF in Coastal Command for reconnaissance, convoy patrol and protection and anti-submarine activity. The aircraft had a very long range when fitted with extra fuel tanks—a useful feature for patrol work and was critical in closing the mid-Atlantic gap at the height of the U-boat menace. Maximum speed was about 300 mph and the ceiling for the Coastal Command versions was about 30 000 feet It was armed with a mixture of 20 mm cannon and .5 and .303 inch machine-guns. Maximum bomb load was 12 800 lbs. The range of the normal bomber version was about 3000 miles but was greatly extended in the Coastal Command versions. (RAF, RAAF, US)

Lightning (P-38) (Lockheed)
A twin-engined twin-tailed fighter with a maximum speed of 414 mph and service ceiling of 43 800 feet. Normal range was 500 miles but could be extended to 1910 miles. Armament was one 20 mm cannon and four .5 inch machine-guns. (US)

Lysander (Westland)
The Lysander was a single-engined high wing monoplane built before the war as an army cooperation aircraft. It was built to take off from and land on very short strips and could fly at very slow speed. It had a maximum speed of 224 mph and a ceiling of 23 000 feet. It was originally designed to be armed with three .303 inch machine-guns and to carry 560 lbs of bombs but was probably never used in its original capacity. Range was 590 miles. (RAF)

Macchi C 200 (Macchi)
A single-engined low-wing monoplane used as a fighter. It was armed with two 12.7 mm machine-guns. Maximum speed was 310 mph and service ceiling was 33 000 feet. Range was 1000 miles. (Italian)

Marauder (Martin) (B-26)
A twin-engined monoplane used as a bomber or torpedo bomber. Maximum speed of about 305 mph and ceiling of 32 500 feet. Armed with anything from five to twelve .5 inch machine-guns and could carry 5200 lbs of bombs. It had a range of 1800 miles. A crew of six or seven. (US and RAF)

Maryland (Martin)
A twin-engined monoplane used as a bomber or torpedo bomber. Maximum speed was 278 mph and service ceiling was 28 000 feet. Armament was eight .303 inch machine-guns and it could carry up to 2000 lbs of bombs. It had a range of 1080 miles. A crew of six or seven. (US and RAF)

Me-109 (Messerschmitt)
A single-engined single-seat fighter, it was the German counterpart of the Spitfire and was used throughout the war. The two aircraft were very similar in performance. Maximum speed varied between about 355 mph and 400 mph with a ceiling of about 38 500 feet. Armament was commonly three 20 mm cannon and two 13 mm machine-guns. Range was about 560 miles. (German)

Me 110 (Messershmitt)
A twin-engined aircraft originally built as two seat fighter but could not compete with the single-seaters. Developed as a night fighter and for ground attack. Maximum speed was 360 mph. Armament was a variable mix of 20 mm cannon and 7.9 mm machine-guns. Maximum range was 1300 miles. The crew was two. (German)

Mitchell (or B-25) (North American)
A twin-engined monoplane bomber. Maximum speed was 272 mph and service ceiling was 25 000 feet. Armament was ten, twelve or fourteen .5 inch machine-guns or one 75 mm cannon and eight .5 inch machine-guns. Range was 1450 miles with a bomb load of 3200 lbs. Crew was five. (RAF and US)

Mosquito (De Havilland)
A twin-engined monoplane used as a fighter, fighter bomber, bomber and photo reconnaissance aircraft, the Mosquito had a remarkable performance and versatility. It was unique in that its main components were constructed from laminated wood. The maximum speed was 417 mph and the ceiling normally about 38 000 feet. In the fighter version it was armed with four .303 inch machine-guns and four 20 mm cannon in the nose. The bomber version was unarmed but could carry up to 4000 lbs of bombs internally. The range with full bomb load was in excess of 1500 miles. The crew was two. (RAF and RAAF)

Mustang (P-51) (North American)
A single-engined low-wing monoplane used as a fighter. In its later versions it proved a valuable escort for the US daylight bombing offensive deep into Germany. It was armed with four .5 inch machine-guns and had maximum speed of 486 mph Its service ceiling was 42 500 feet and its range 2160 miles. As a fighter bomber it could carry two 1000-lb bombs. (US RAF)

Savoia SM.79 (Savoia Marchetti)
A three-engined low-winged monoplane bomber. It had a range of 1190 miles and carried a bombload of 2750 lbs. Armament was three 12.7 mm and two 7.7 mm machine-guns. Maximum speed was 255 mph and the service ceiling was 27 880 feet. Crew was three. (Italian)

Spitfire (Supermarine)

The Spitfire was a single-engined single-seat fighter which formed the backbone of Fighter Command from the end of the Battle of Britain till after the end of the war. It was mainly used in the fighter role but was also used as a fighter bomber. The maximum speeds varied from 355 to 454 mph The ceilings varied from 34 000 to 43 500 feet. Armament consisted of a mixture of 20 mm cannon and .303 inch machine-guns. It could carry a bomb load of 1000lbs. (RAF and RAAF)

Stork (Fieseler)

A single-engined high-wing monoplane. A communications aircraft designed for ferrying one or two passengers. It could fly and land at very low speed which allowed it to operate into very small unprepared fields. Maximum speed was 109 mph and the service ceiling was 17 000 feet. Armed with one 7.9 mm machine-gun. Range about 545 miles (German)

Sunderland (Short)

A four-engined flying boat used by Coastal Command for anti-submarine patrols, maritime reconnaissance and convoy protection. Maximum speed was 247 mph and the service ceiling was 17 200 feet. Armed with eight to twelve .303 inch machine-guns. Could carry up to 2000 lbs of bombs or depth charges. Maximum range was 2530 miles. Crew ten or eleven. (RAF and RAAF)

Swordfish (Fairey)

A single-engined biplane developed well before the war for use from aircraft carriers. It was used for anti-submarine patrol, reconnaissance and as a bomber. Maximum speed was 139 mph and the operational ceiling was 17 000 feet. Armed with two .303 inch machine-guns. Maximum range was 528 miles with 1500 lbs. of bombs. Crew of two or three. (Fleet Air Arm)

Tempest (Hawker)

A single-engined single-seat fighter. It achieved fame as the destroyer of the German flying bombs. Maximum speed was 457 mph and ceiling 38 800 feet. It was armed with four or six 20 mm cannon or two 20 mm cannon and four .5 inch machine-guns and could carry up to 2000 lbs of bombs. Range with drop tanks was up to 1640 miles. (RAF)

Thunderbolt (Republic)

A single-engined single-seat fighter used by a number of air forces as a fighter and fighter bomber. It had a maximum speed of 470 mph and a ceiling of 42 000 feet. It was armed with up to eight .5 inch machine-guns and could carry up to 2000 lbs of bombs. Range was 1970 miles. (US and RAF)

Tiger Moth (De Havilland)

A single-engined biplane primary training aircraft which was used very widely for initial pilot training throughout the Empire Air Training Scheme. A very nice little aircraft to fly. Maximum speed 109 mph and service ceiling 14 000 feet. Range was 258 miles. It carried no armament and no bomb load. (RAF, RAAF, RNZAF and RCAF)

Tomahawk (or P-40) (Curtiss)

A single-engined single-seat fighter, this was the first of the P-40 series of fighters. It had a maximum speed of 340 mph with a ceiling of 30 000 feet. It was armed with two .5 inch machine-guns and two .303 inch machine-guns (RAF aircraft had the .5 inch replaced by .303 inch). The maximum range was 890 miles. (US and RAF)

Typhoon (Hawker)

A single-engined single-seat fighter introduced as the replacement of the Hurricane. Its maximum speed was 405 mph and its ceiling 34 000 feet. It was armed with four 20 mm cannon and could carry up to 1000 lbs of bombs or eight 60-lb or 25-lb rockets. Maximum range was up to 1000 miles. (RAF)

Valentia (Vickers-Armstrong)

An open cockpit biplane which was virtually obsolete at outbreak of war. It had a maximum speed of 110 mph and a range of 770 miles. Its use was restricted to some training (e.g., wireless operators) and some communications work.

Vengeance (Consolidated-Vultee)

A single-engined low-wing monoplane used as a dive-bomber. Maximum speed was 260 mph and the service ceiling was 22 000 feet. It could carry 1500 lbs. of bombs over a range of 700 miles. Crew was two. (RAF, RAAF and US)

Ventura (Lockheed)

A twin-engined monoplane bomber. Maximum speed was 298 mph and service ceiling was 25 000 feet. Armed with seven .5 inch and two .3 inch machine-guns, it could carry 5000 lbs. of bombs and had a range of 1375 miles. Crew of four. (RAF and RAAF)

Walrus (or Seagull) (Supermarine)

A single-engined biplane amphibian flying boat. Used for reconnaissance and air sea rescue. A very rugged machine which could be catapult-launched from ships and landed in the open sea for recovery. Maximum speed 135 mph and ceiling 16 800 feet. Armed with two .303 inch machine-guns and could carry 500 lbs of bombs. Maximum range 600 miles. Crew of two to four. (RAF and RAAF)

Warwick (Vickers–Armstrong)

A twin-engined low-wing monoplane, the Warwick was developed as a larger version of the Wellington. It was beaten into action by the four-engined bombers and was used in air sea rescue, transport and some general reconnaissance. (RAF)

Wellington (Vickers)

A two-engined aircraft used as a bomber, torpedo bomber, anti-submarine aircraft and for general reconnaissance and training aircraft at OTUs. The Wellington carried out the first bombing attack of the war by the RAF and served right through to the end of the war. It was built in a number of versions. Maximum speed varied from 235 to 255 mph and ceiling from 18 000 to 19 500 feet (a version with a pressurised cabin could reach 36 700 feet). Armed with six to eight .303 inch machine-guns it could carry up to 8000 lbs of bombs or two torpedoes. As an anti-submarine aircraft it was fitted with the Leigh searchlight and carried depth charges. Crew of five or six. (RAF)

Wirraway (Commonwealth Aircraft Corporation)

A single-engined low-wing monoplane built in Australia as an adaptation of the Harvard. It had been intended as a general-purpose aircraft but disastrous meetings with the Japanese Zero led to its immediate withdrawal to training and some army cooperation duties. It had a maximum speed of 300 mph. Basic armament was two 7.7 mm machine-guns and two 20 mm cannon. Range was 1100 miles and service ceiling was 32 000 feet. (RAAF)

Zero (Mitsubishi)

A single-engined single-seat fighter or fighter bomber. The Zero was a very agile fighter which took the Allies by surprise when they first met it. Maximum speed was 350 mph It was armed with two 7.7 mm machine-guns in the nose and two 20 mm cannon in the wings. The maximum range was about 900 miles. (Japanese)

BIBLIOGRAPHY

ALLEN, Lois, *Burma The Longest War*, J.M. Dent and Sons Ltd, London, 1984.

BIDWELL, Shelford, *The Chindit War*, Hodder and Stoughton, London, 1979.

BRUNE, Peter, *Gona's Gone*, Allen and Unwin, St Leonards NSW, 1994.

BUCKHEIM, Lothar-Gunther, *U-Boat War*, Collins, London, 1975.

CALLAHAN, Raymond, *Burma 1942-45: The Politics and Strategy of the Second World War 1942-45*, Davis Poynter, London, 1978.

CHURCHILL, W.S., *The Second World War*, Cassell, London, 1964.

FOOT, M.R., *SOE: The Special Operations Executive*, BBC, London, 1984.

HERRINGTON, J., *Air War Against Germany and Italy 1939-1943*, Australian War Memorial, Canberra.

HOYT, Edwin P., *The Death of the U-Boats*, McGraw Hill, New York, 1988.

INNES, David, *Beaufighters over Burma*, Blandford Press, Poole, Dorset, 1985.

JAMES, Richard Rhodes, *Chindit*, John Murray, London, 1980.

Jane's Fighting Aircraft of World War II, Foreword by Bill Gunston, Military Press, New York.

KEMP, Paul, *The Russian Convoys 1941-1945*, Arms and Armour Press, Poole, Dorset, 1987.

KIRBY, Woodburn S., *The War Against Japan*, HMSO, London, 1965 ·

LOCKWOOD, Douglas, *Australia's Pearl Harbour Darwin 1942*, Cassell Melbourne, 1966.

McAULAY, Lex, *To The Bitter End*, Random House, Milson's Point, 1992.

McCARTHY, Dudley, *South-West Pacific Area—First Year*, Australian War Memorial, Canberra, 1959.

MACINTYRE, Donald, *The Battle of the Atlantic*, Severn House, London, 1975.

MIDDLEBROOK, Martin, *Convoy*, William Morrow, New York, 1976.

ODGERS, George, *Air War Against Japan 1943-45*, Australian War Memorial, Canberra, 1968.

ROBERTSON, J., *Australia at War*, William Heinemann, Melbourne, 1981.

ROONEY, David, *Burma Victory*, Arms and Armour Press, Poole, Dorset, 1992.

ROSKILL, Stephen, *Churchill and the Admirals*, William Collins, London, 1977.

SCHOFIELD, B.R., *The Russian Convoys*, B.T. Batsford, London, 1964.

SLIM, Sir William, *Defeat into Victory*, Cassell, London, 1956.

WALKER, Allan, *The Island Campaigns*, Australian War Memorial, Canberra, 1962.

WALKER, Brian, *Black Jack*, Banner Books, Belconnen, ACT, 1994.

PHOTOGRAPHIC ACKNOWLEDGMENTS

Key t: top; b: bottom; l: left; r: right

Page 9 Doug Moffat; **11** Jim Savage; **14** Doug Moffat; **18** Reg Torrington; **19** (t) Reg Torrington; (b) Reg Hackshall; **22** John Appleton; **27** Jim Savage; **29** Doug Moffat; **31-44** Jack Stronach; **45-8** Jack Robertson; **49** Jack Stronach; **50** Jack Robertson; **51** Ron Turner; **52** Jack Stronach; **54** Gerry Judd; **55** Ross Pearson; **59-62** Jack Stronach; **63** Laurie Jones; **68** Doug Nicol; **71** RAAF Association; **76-77** Jack Stronach; **81** David Innes; **82** (l) Ross Pearson; (r) Bill Hughes; **85** Ross Pearson; **89** Reg Hackshall; **92** Dennis Hornsey Collection; **93** Ron Barker; **96** (t) Leo Allen; (b) Bill Hughes; **100** Fred Cassidy; **101** Bruce Edenborough; **102** Fred Cassidy; **103** Reg Hackshall; **107** Ron Barker; **108** Fred Cassidy; **110** Leo Allen; **112** Fred Cassidy; **115** Bob Hartman; **116-7** Bruce Edenborough; **124** (l) RAAF Association (r) Jack Stronach; **126** Jack Stronach; **132-6** Syd Kildea; **142** Dennis Hornsey Collection

INDEX